# ABANDONED ASYLUMS AND **INSTITUTIONS** OF **NEW YORK**

## THE EMPIRE STATE STRIKES BACK

DAVE SNOOK

—
AMERICA
—
THROUGH
—
TIME

America Through Time
Fonthill Media Inc.
www.through-time.com

First published 2025
Copyright © Dave Snook 2025

ISBN 978-1-63499-503-0

All rights reserved. No part of this publication may be reproduced, stored in a retrieval system or transmitted in any form or by any means, electronic, mechanical, photocopying, recording or otherwise, without prior permission in writing from Fonthill Media Inc.

Typeset in Trade Gothic 10pt on 15pt
Printed and bound in England

# ACKNOWLEDGMENTS

I would like to acknowledge all the people with whom I have explored over the years to make this project possible. Although there are too many to count, I wanted to extend a special thanks to a few individuals: Jason Baker, Robert Duffy, Cara Duplessis, Jeremy Folcik, Brett Kane, Michael Martin, Cole Merlot, Scott Militello, Timothy Neary, Bill Stephens, Isaac Wegerson, and Becca and Rich Whiddon. Here's to all the late-night drives, pre-dawns, long walks, runs, countless hours on the road, chats, and laughs.

-Thank You

# CONTENTS

Acknowledgments          **5**

Introduction          **7**

1   Exclusion          **9**

2   Medical          **21**

3   Plans          **37**

4   Infectious          **65**

5   Recreation          **75**

6   Eugenics          **85**

About the Author          **96**

# INTRODUCTION

This book is not just a collection of photographs, but a journey through New York's asylums and institutions. It delves into the exclusion and medical practices of the mentally ill, architectural plans, infectious diseases, recreation, and the eugenics movement, all captured through my lens. Each place I visit has a unique history, energy, personality, and beauty, which I am eager to share with you. But before we embark on this journey, let me share with you the two primary reasons that sparked my passion for this photographic exploration.

The first reason is that I was deeply intrigued by abandoned state hospitals I have visited over the years. I have always found their grandeur, size, and history fascinating. What baffled me the most was why these architectural marvels, which hold such rich American history, were left to decay. These asylums were once self-sufficient towns away from society, but now they are left to rot and be forgotten. I aimed to capture and document as much as possible before these places became distant memories. Documenting these buildings inside and out is essential to keep their history alive for future generations. Most American asylums have a troubling history of abuse, causing many states to look the other way and neglect preservation efforts, making preserving and saving these historical buildings not easy. Most are demolished before any action, funds, or awareness is made. That's why I started documenting these abandoned historical asylums.

The second reason is my belief in the transformative power of curiosity. As humans, we are innately curious about the unknown. We have an insatiable desire to unravel the mysteries of the world around us and to discover new ideas and places. For me, there is something inexplicably mesmerizing about these abandoned structures. Once filled with life and energy, they now sit silently, relinquishing themselves to the elements. The scenes both inside and outside these structures evoke a sense of otherworldliness and satisfaction. The death of humankind's artistic imagination is contrasted by nature's sheer will to survive, and it is this very thing that captivates me. As the buildings crumble and decay, new life emerges from every corner, bringing on a unique yet beautiful scene. This book is not just about documenting the past but about inspiring hope for the future. It's about showing that even in the face of decay, there is beauty and potential for renewal.

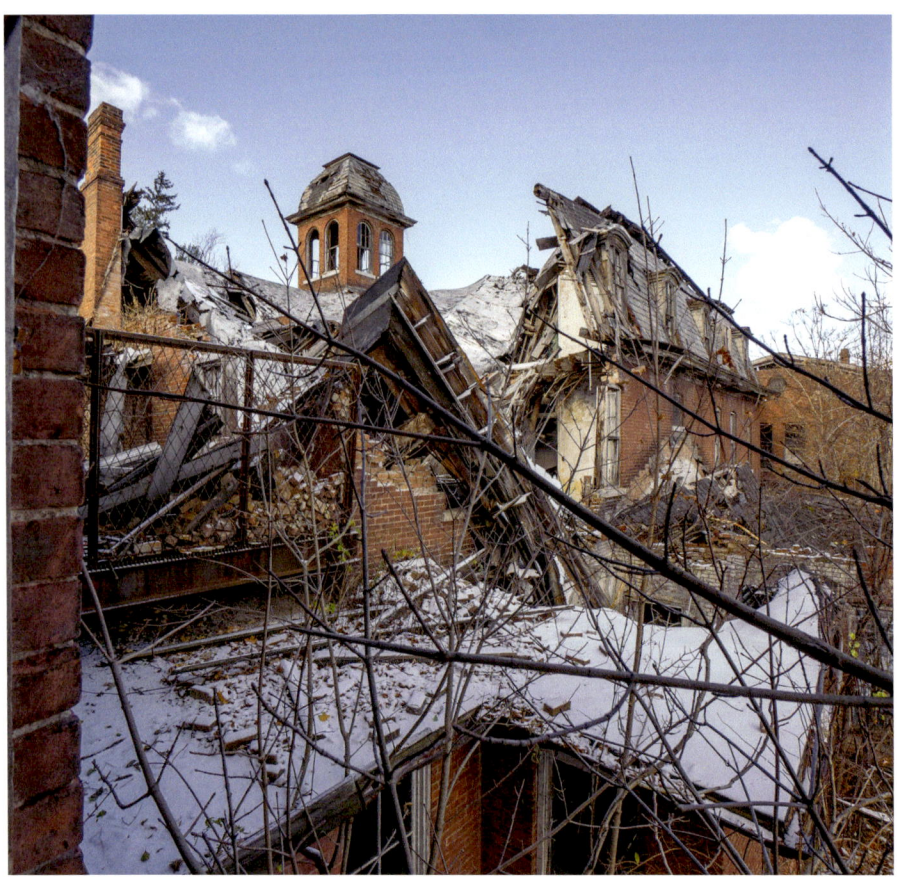

"Do. Or do not. There is no try."

# 1

# EXCLUSION

Since the beginning of the United States, the insane have been excluded from society. Initially, they were sent to almshouses for the poor, as there were no asylums yet for the insane. In 1791, the state of New York opened its first facility as a general hospital with a mental health ward to provide care for insanity. However, it was not until the growing number of insanity cases that the board of governors requested the legislature to construct a separate building for this class of patients. The legislature granted an annual sum of $12,500 for fifty years. In 1808, the new facility for the insane and sick persons was complete, under the appropriation that the hospital was of great public utility and humanity. However, there were still no requirements for admission of the insane. In 1809, a law was passed for overseers of a city or town to contract with the governors of New York Hospital for the care of insane persons. The cost of the care would be paid by the city or town the patient was from. Dr. Sylvester Willard, the general surgeon and secretary of the state medical society, was concerned about the overcrowding and the harsh living conditions in the poorhouses. He asked the legislature to pass a law called "An Act Concerning Lunatics." It stated:

> No lunatic shall be confined in any prison, or house of correction, or confined in the same room with any person charged with or convicted of any criminal offense. But he shall be sent to the asylum in New York, or to the county poorhouse or almshouse, or other place provided for the reception of lunatics by the county superintendent (of the poor). If such person is not possessed of sufficient property to maintain himself it shall be the duty of the father and mother, and the children of such person, being of sufficient ability, to provide a suitable place for his confinement and to confine and maintain him in such manner as shall be approved by the overseers of the poor of the city or town.

In the 1830s, with the population of New York reaching almost 2,000,000, it became clear that a new approach was required for the treatment of the insane. The New York General Hospital was overwhelmed and could no longer manage the growing number of patients. As a result, the New York Lunatic Asylum was established in Utica in 1843 to provide care for acute patients. If a patient were deemed incurable after two years of treatment, they would be sent back to their county poorhouse. It was not until 1865 that the Willard Asylum Act was passed, which mandated the removal of all diagnoses of insanity from poorhouses in New York. This move was intended to ensure that those with chronic mental illness were no longer housed in poorhouses but instead were sent to a new state-run institution. This new institution was built as a self-contained and self-sufficient town within a town. It would be known as the New York Asylum for the Chronic Insane.

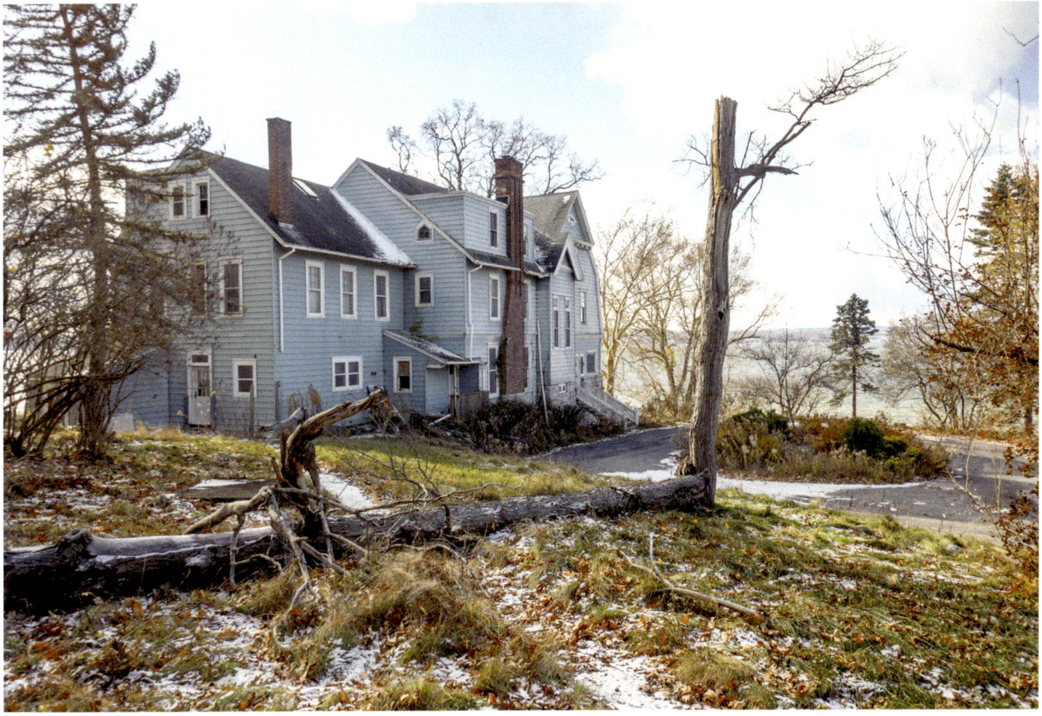

It was typical for acting superintendents to live on the premises or inside the institutions where they worked.

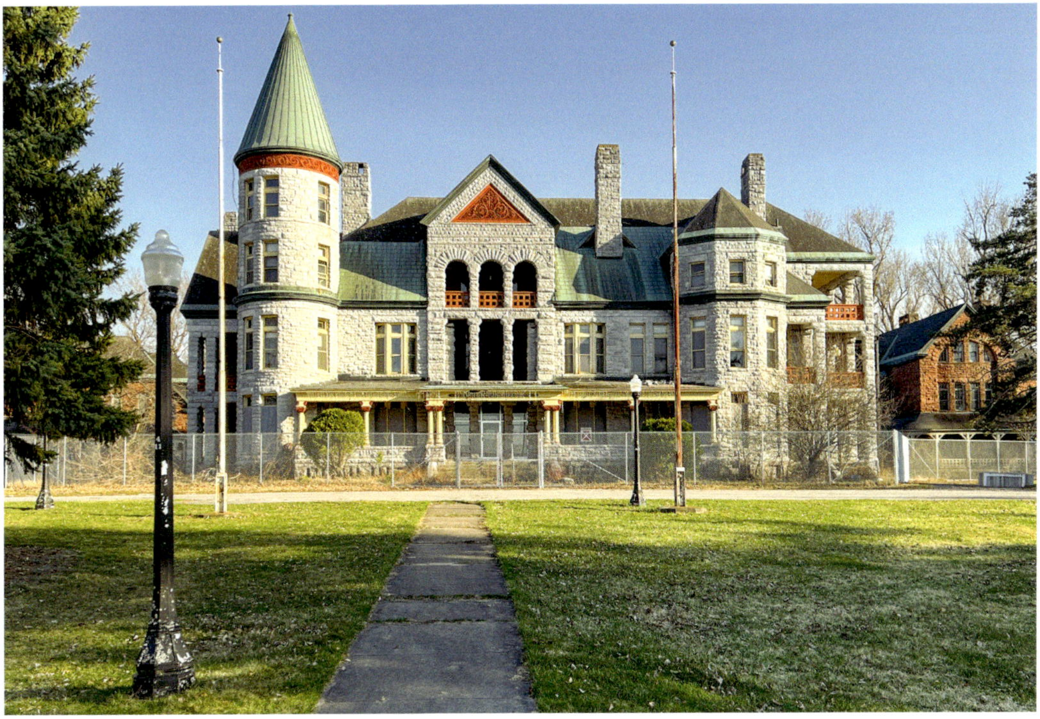

The administration building oversaw daily activities including hiring staff, managing budgets, and meeting with doctors and nurses about schedules and priorities.

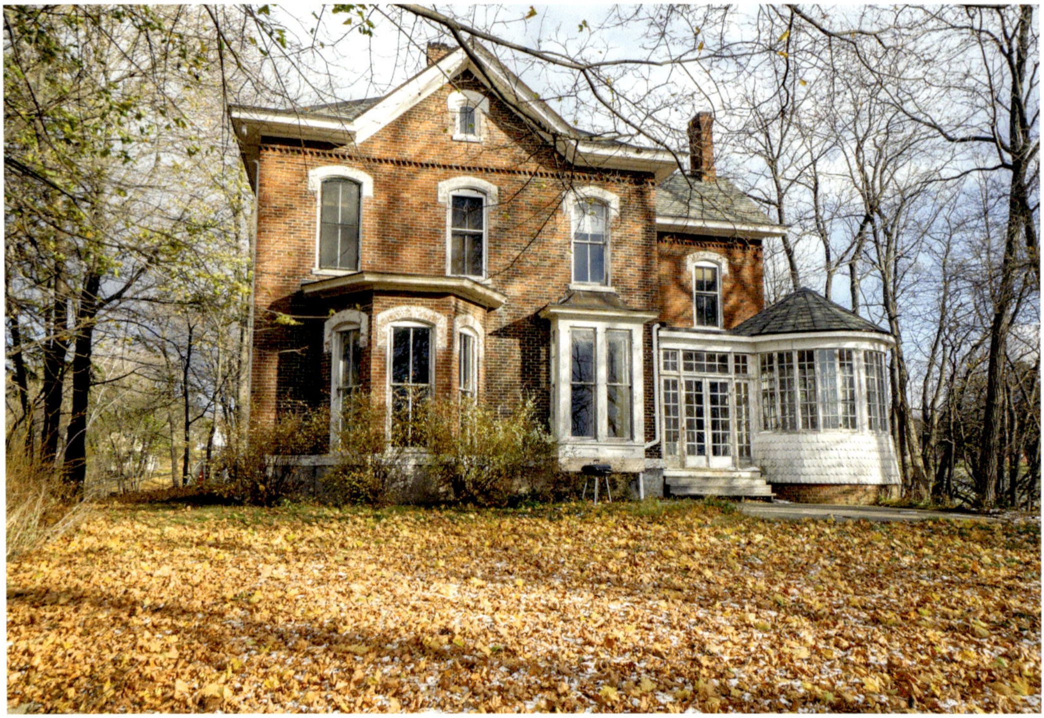

It was important to have housing for the steward and matron who oversaw the daily operations of the asylum.

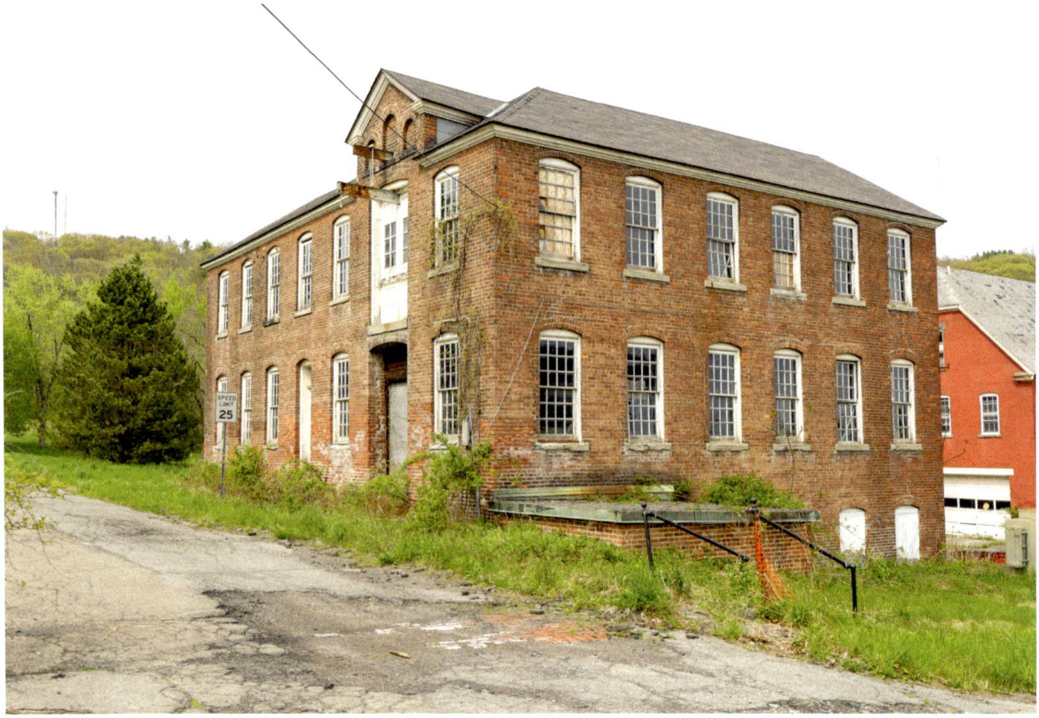

The carpenter's shop maintained, built, fixed, installed, and assembled anything that the asylum needed.

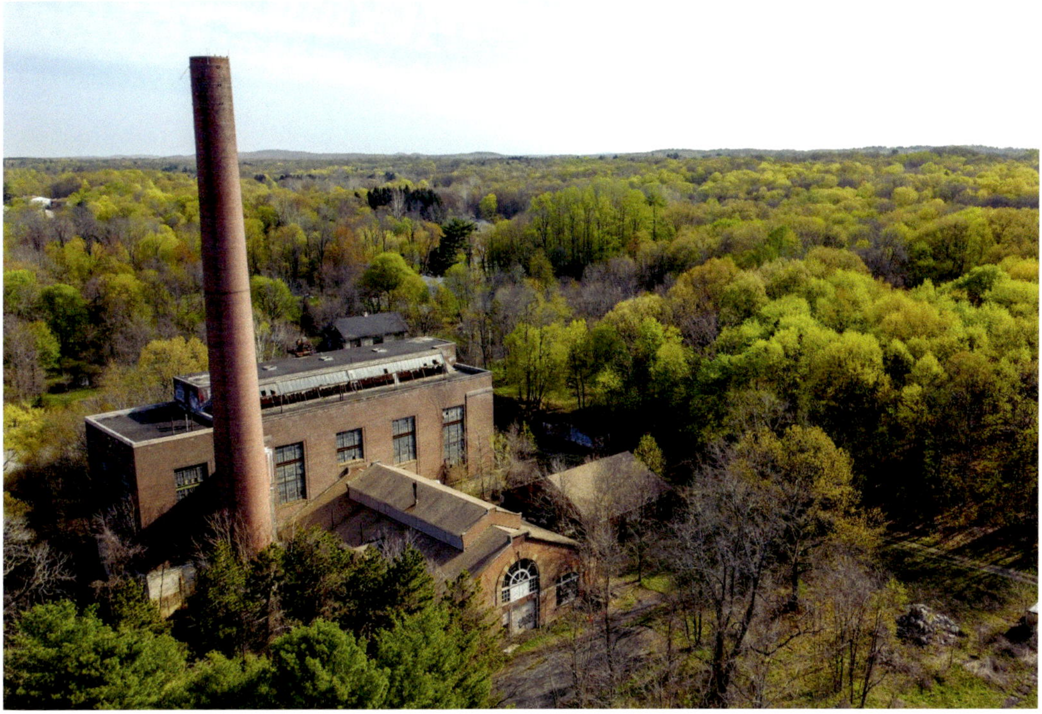

Asylum power plants transformed raw energy sources such as coal into electricity, providing power for the entire institution.

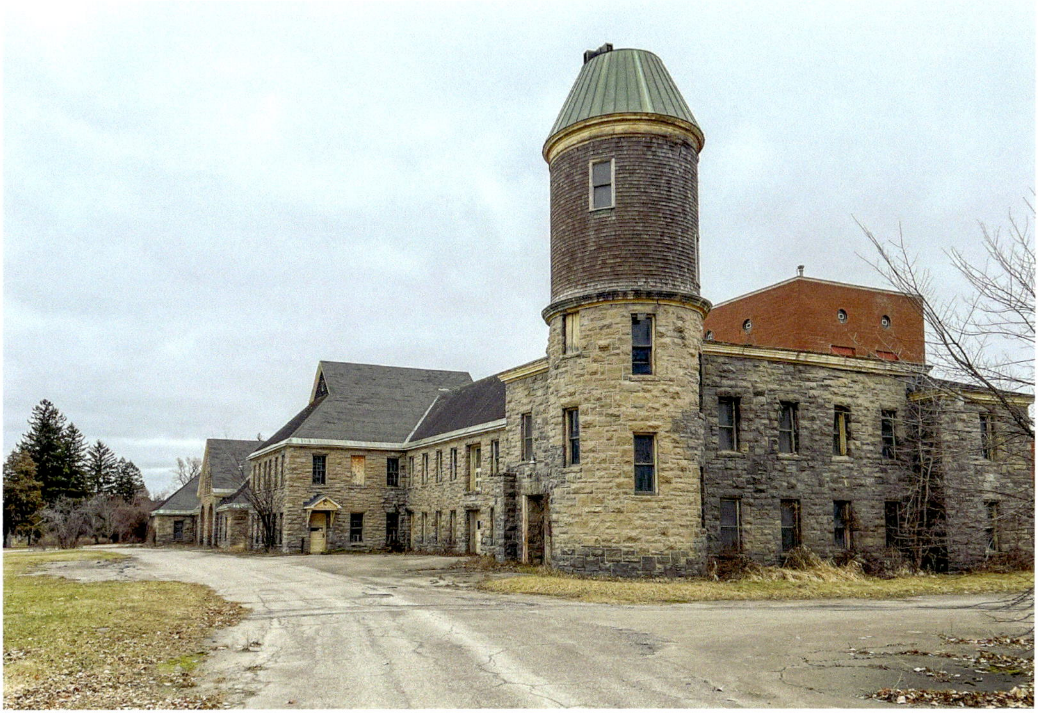

Because of the secluded locations of asylums, a fire station on campus was essential due to the high risk of fires.

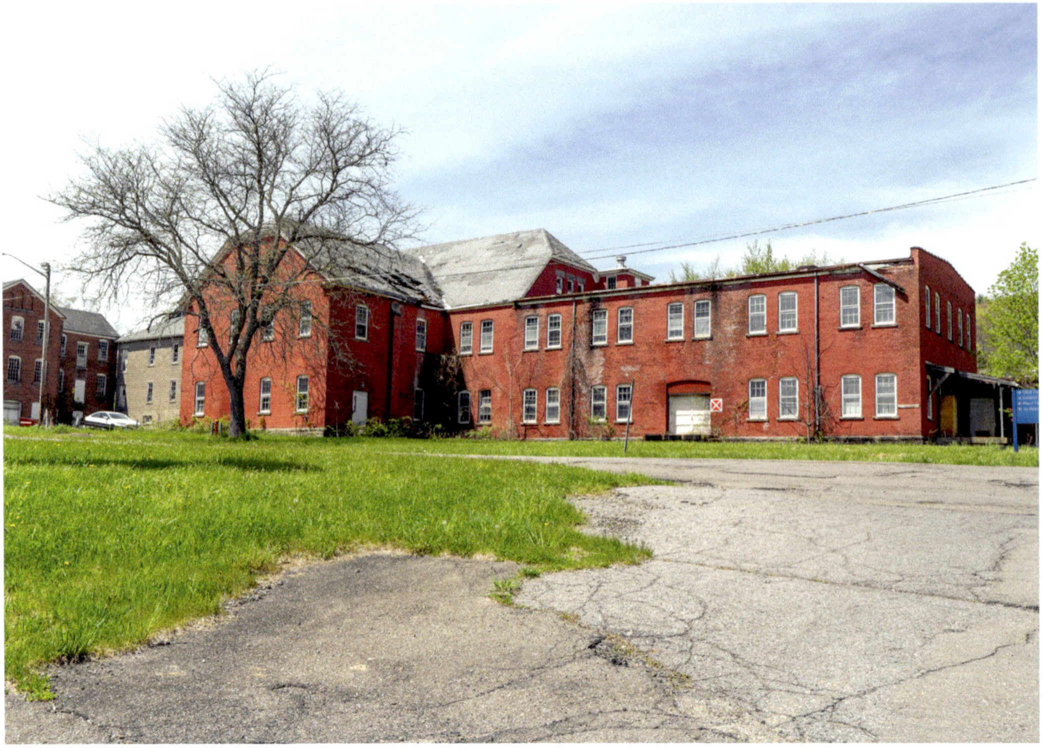

A laundry department was created to keep patients and staff with fresh, clean clothes and linen.

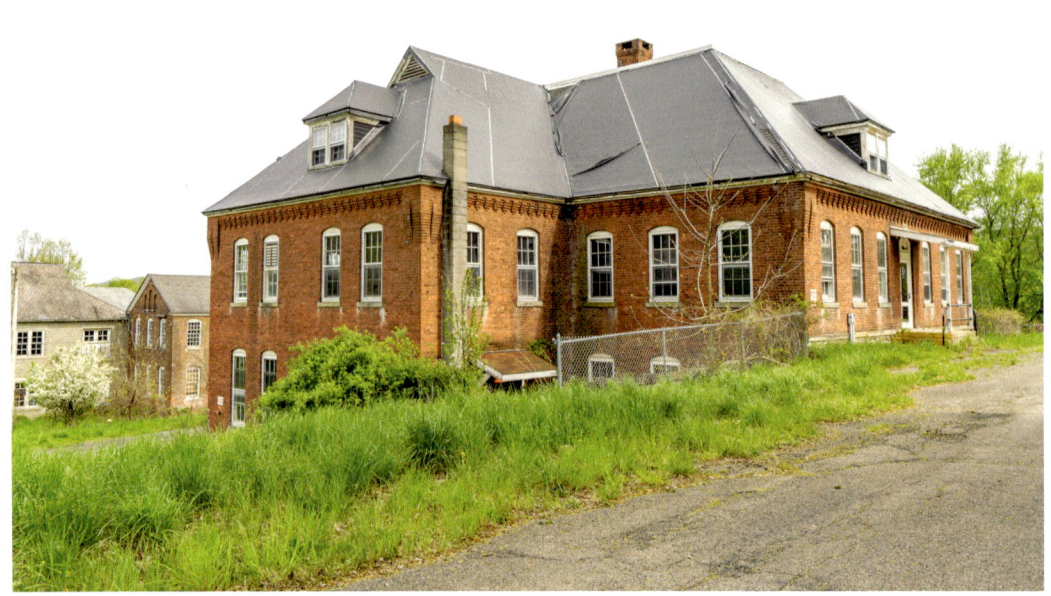

This industrial building was used to manufacture goods for the asylum.

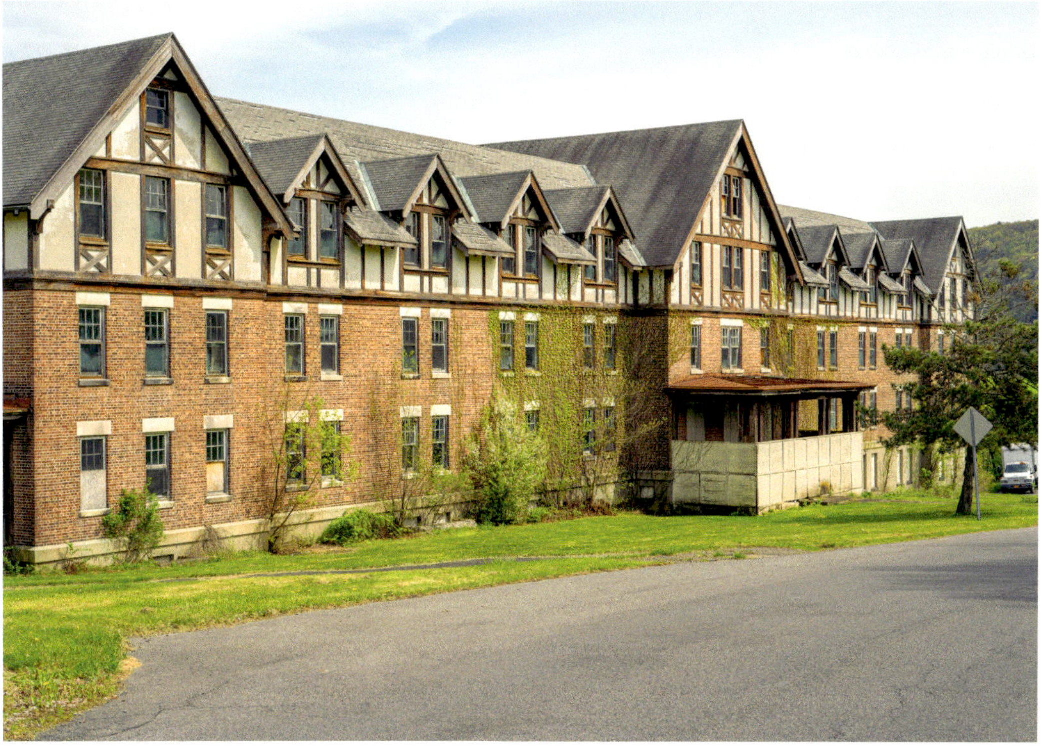

The dormitories were segregated by gender, job titles, and marital status.

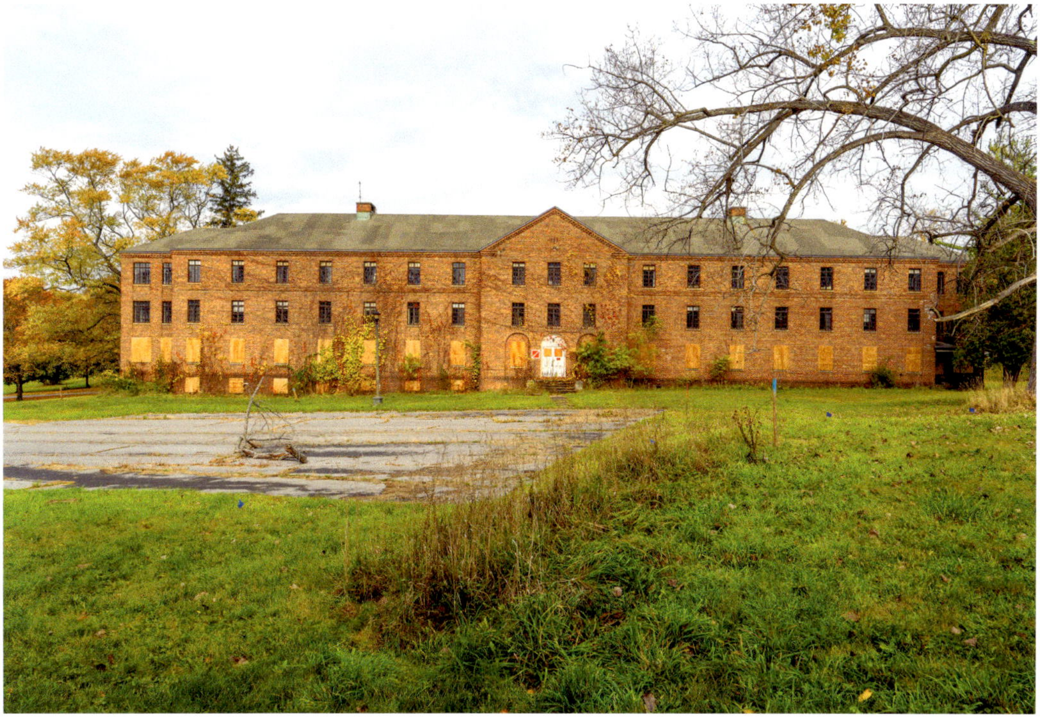

Housing was offered to nurses, staff, married attendants, superintendents, doctors, students, and certain patients.

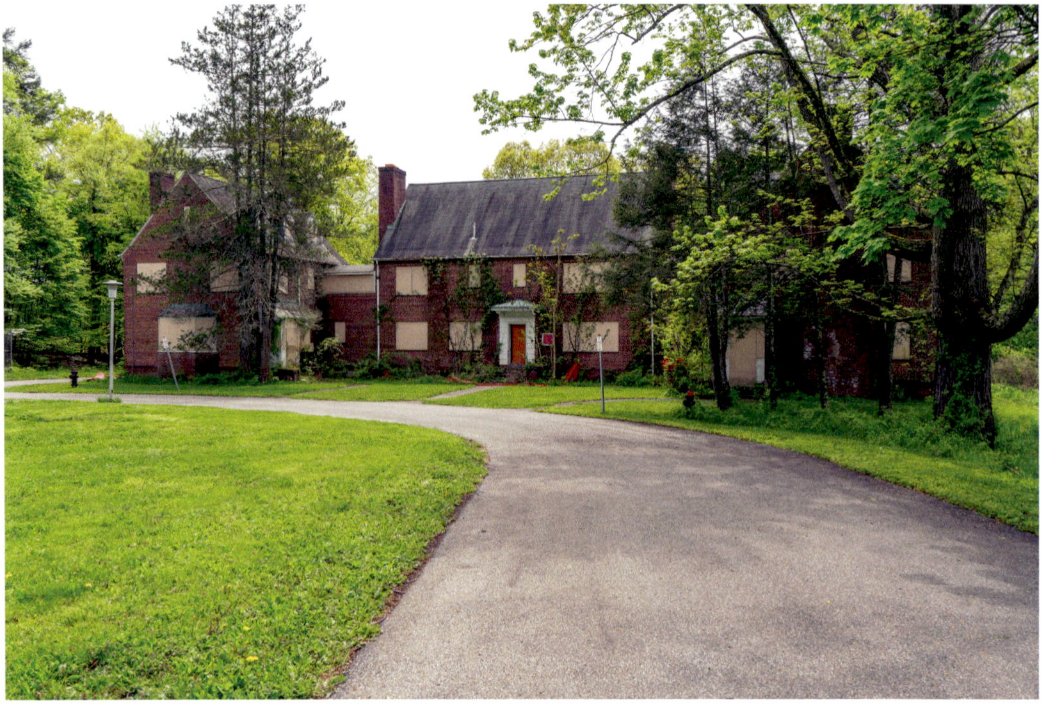

Some institutions not only offered dormitories for staff and patient's but also townhouses on campus.

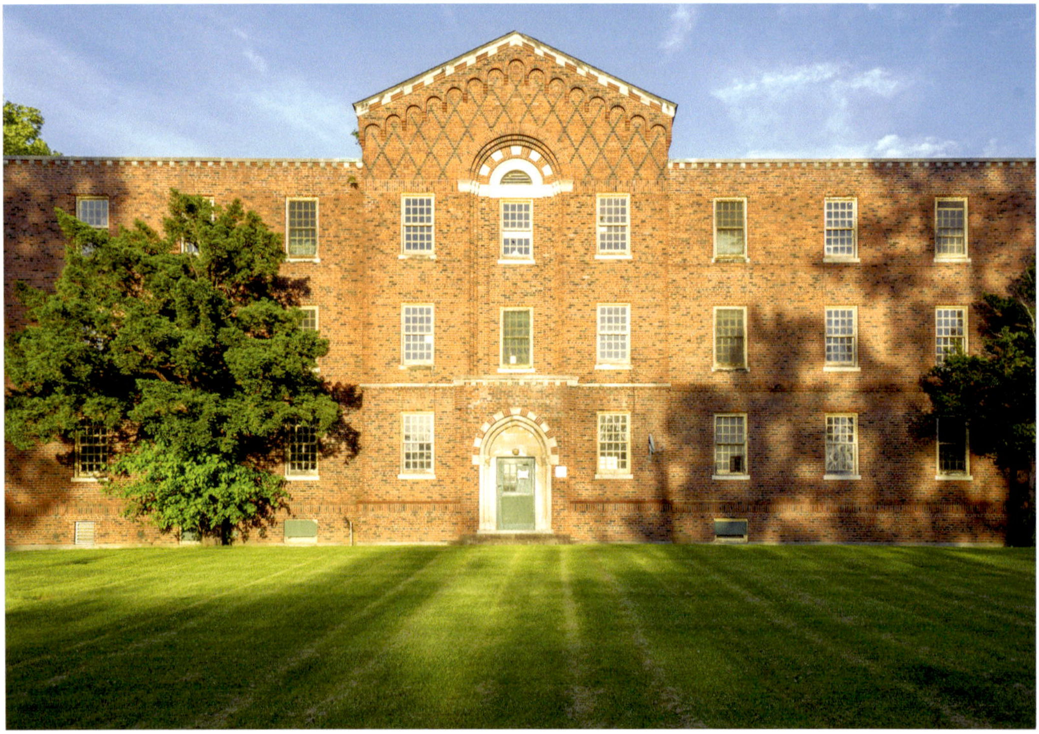

Staff housing on the property was necessary for many reasons, one being that early institutions were built far away from cities, and commuting at the time was difficult and problematic.

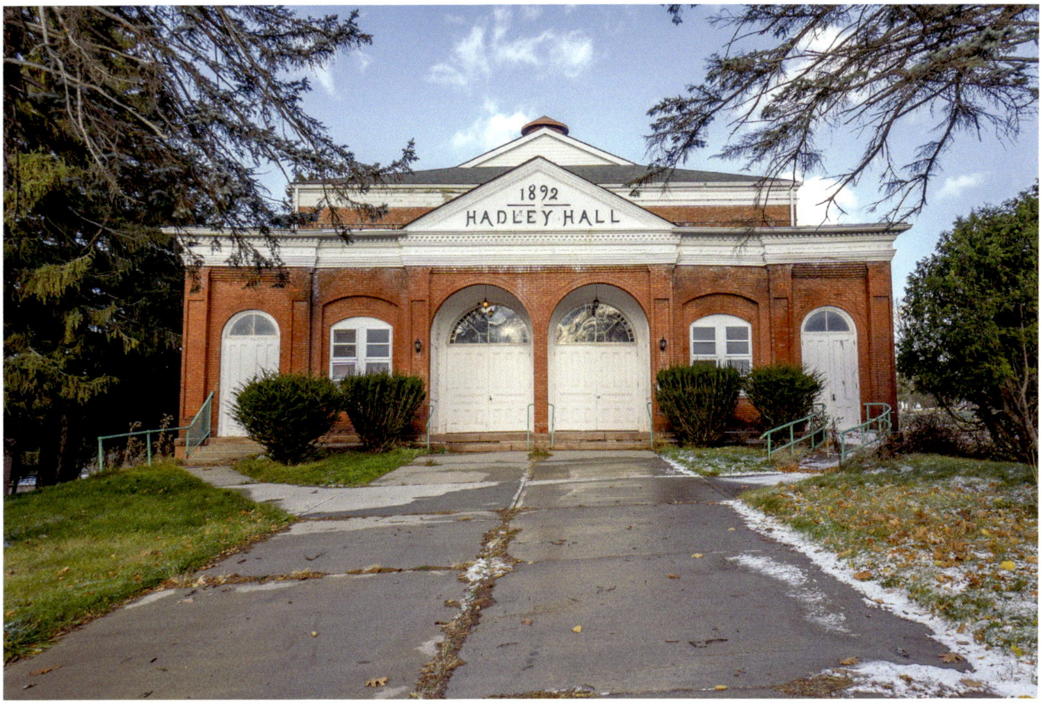

Assembly halls were built to provide patients and staff with fun and entertainment.

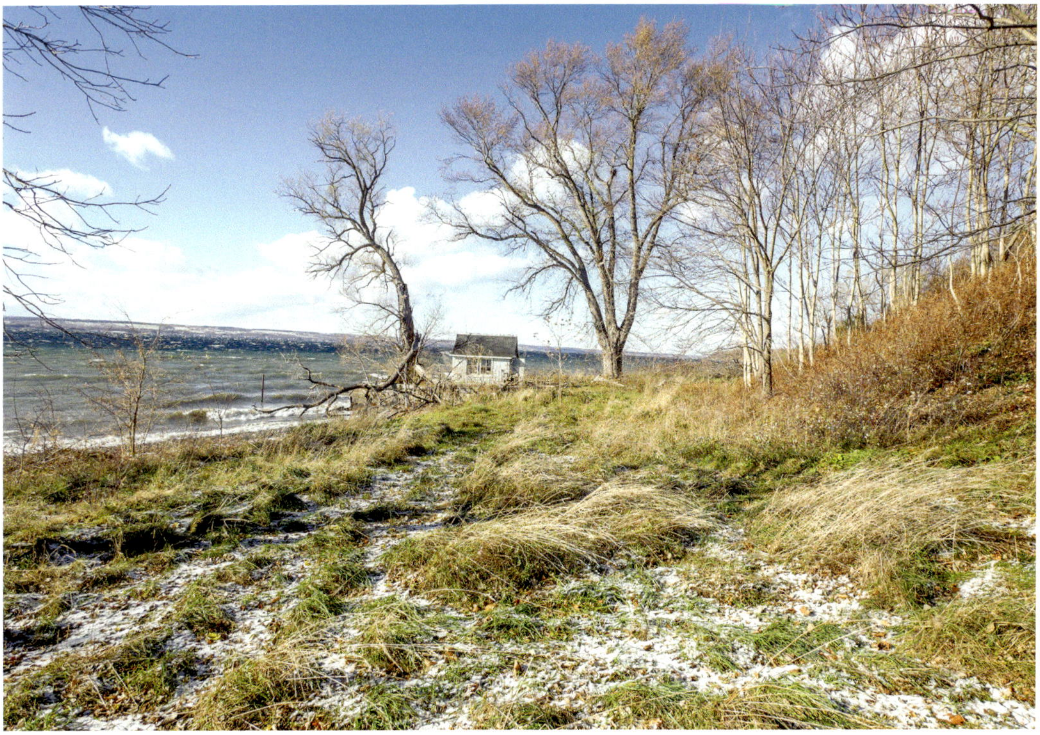

Asylums were often constructed near lakes or rivers because they were easily accessible, and the surrounding natural scenery was considered therapeutic.

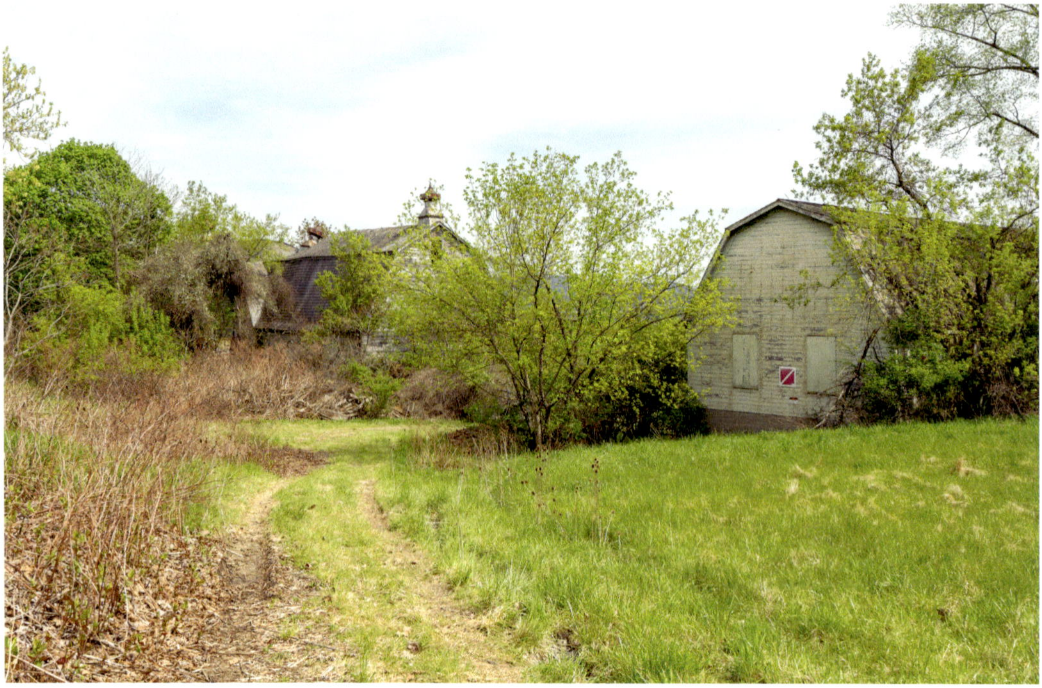

The farm was instrumental in providing patients with opportunities to develop work skills, find purpose, and grow food for the hospital.

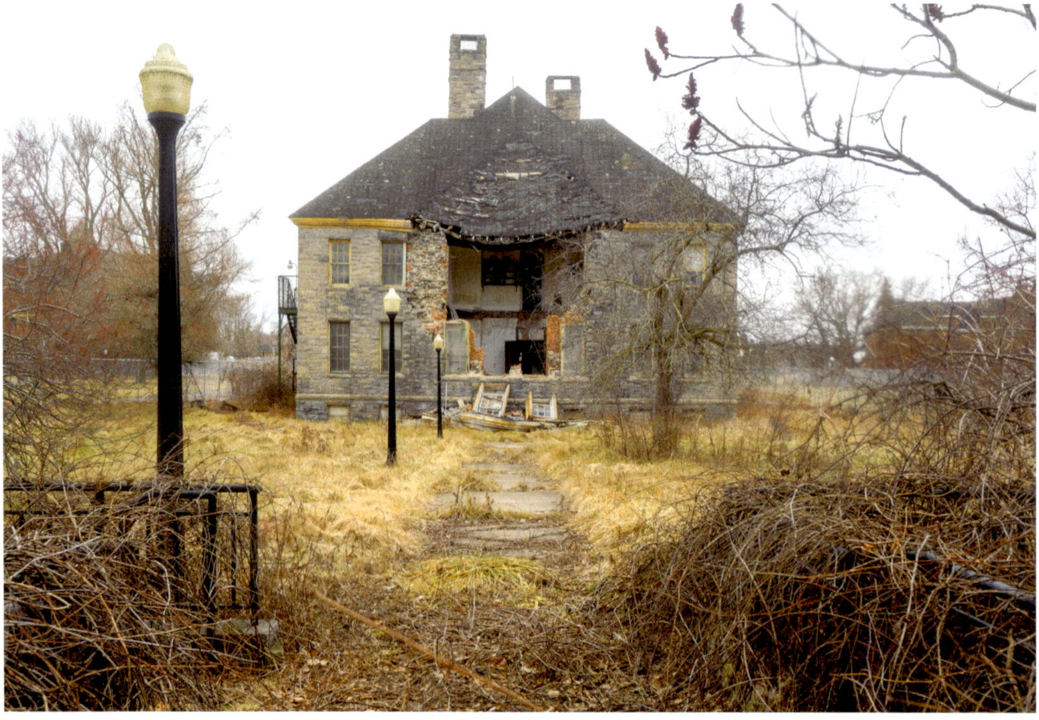

A food service building would help distribute food from the farms to the kitchens for staff and patients to enjoy.

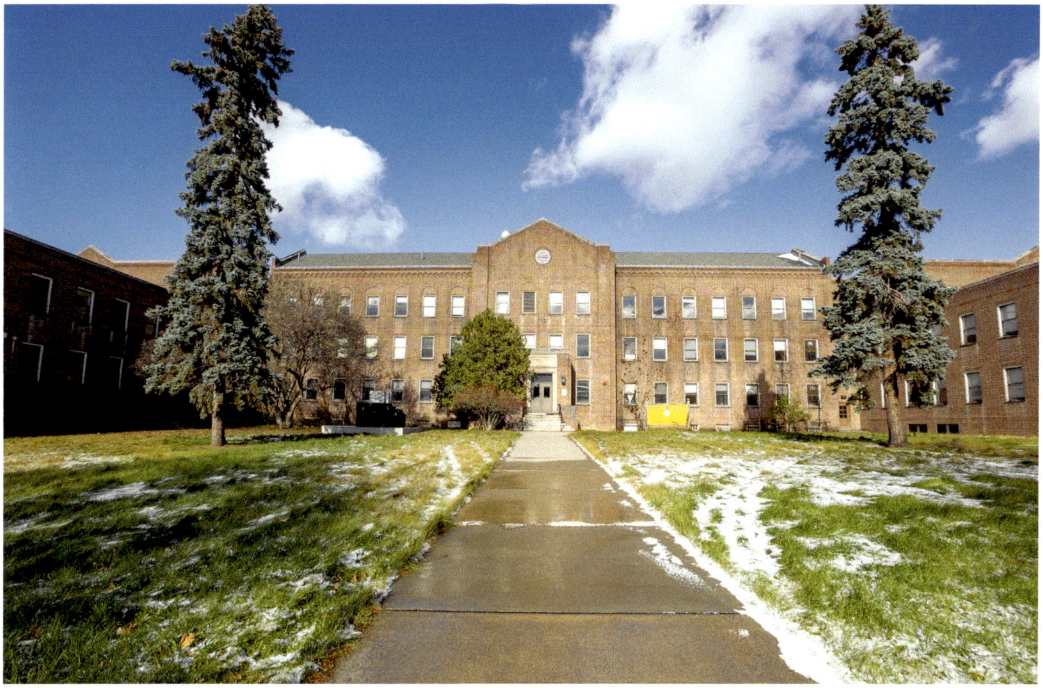

A medical and surgical building was needed to accommodate doctor's offices, dental clinics, outpatient services, diagnostic centers, surgeries, and therapy.

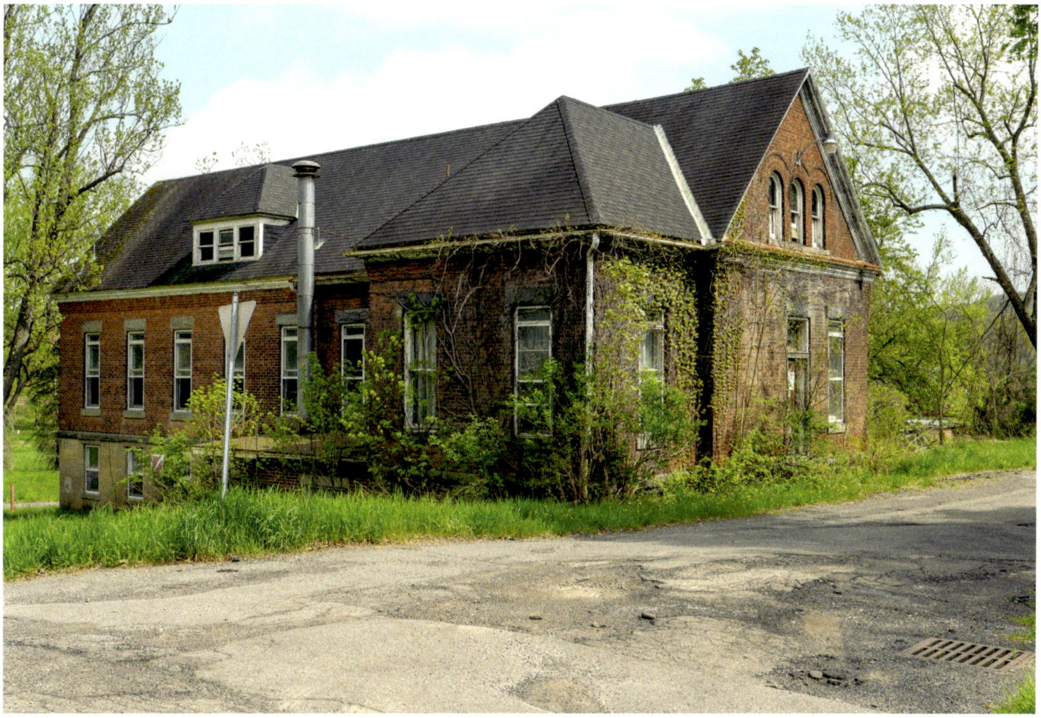

A morgue and laboratory are essential for determining the cause of death, ensuring proper burial, and conducting research on the deceased.

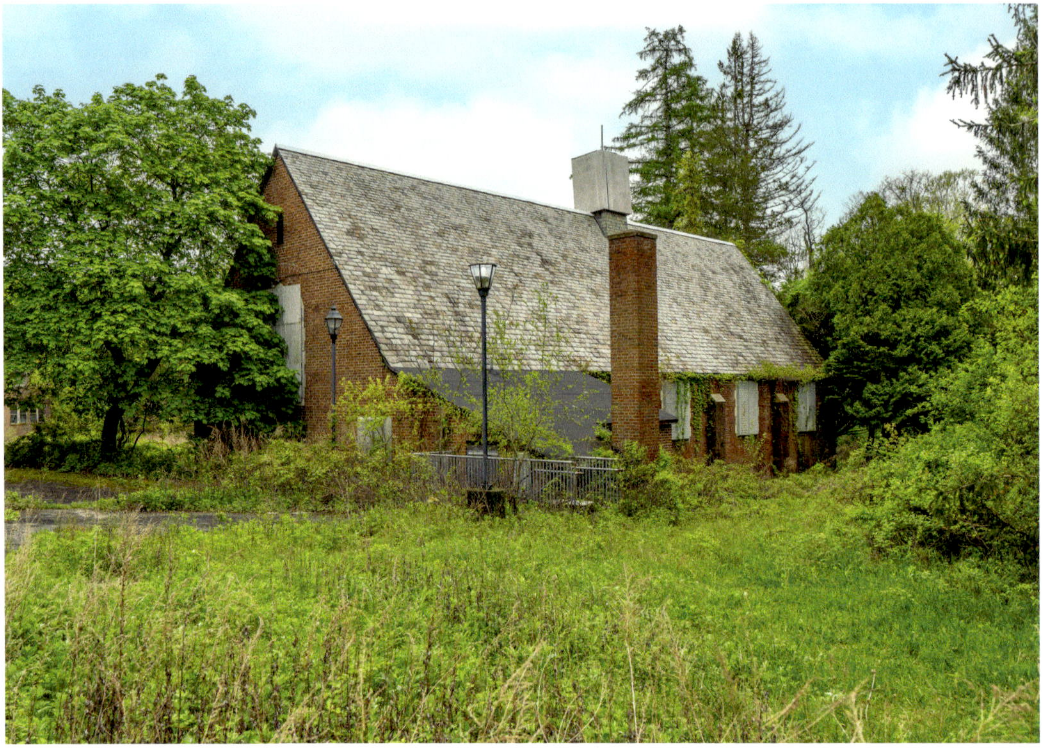

Patients were provided with churches and chapels for religious services to assist them in reflecting and connecting.

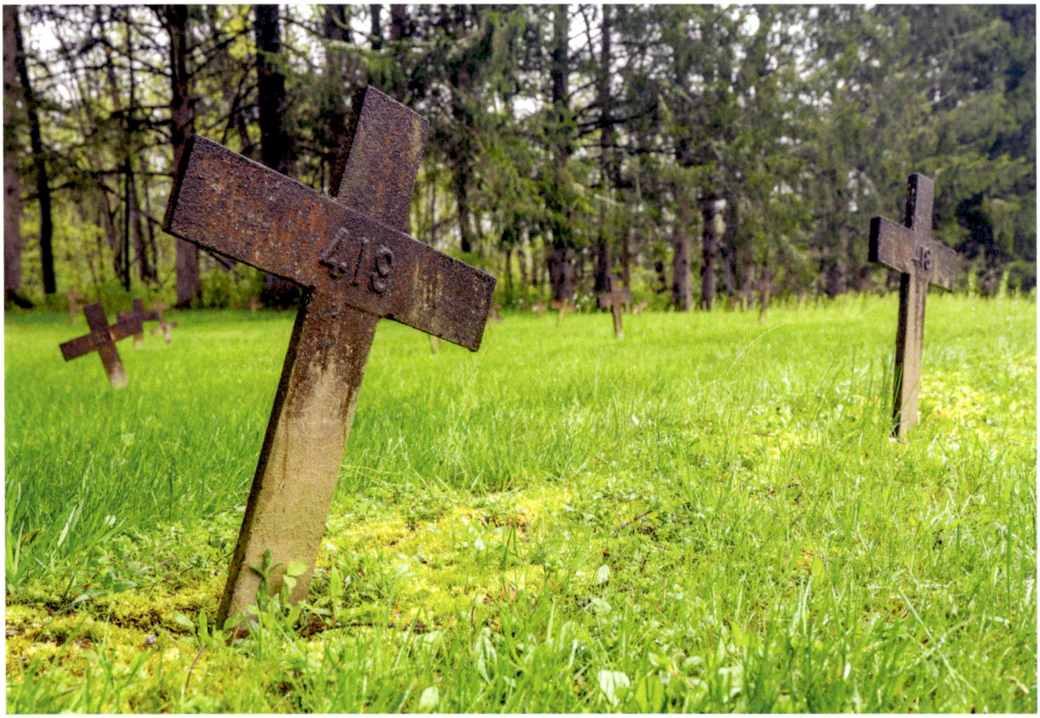

Graveyards were usually on the property or in local plots within the town of the asylum.

# 2
# MEDICAL

The medical treatment of mental illness in the past may appear to be inhumane by modern standards; however, many prominent doctors of that era endorsed such therapies. In the absence of any available medicines or pills to "cure" mental illness, alternate methods were used for treatment. With every medical advancement and new theory, the treatment of mental illness has undergone significant changes.

During the early years of asylums, patients were often restrained with mechanical devices such as straitjackets, manacles, waistcoats, and leather wristlets. These restraints were used for hours or even days at a time. Although doctors claimed that these restraints were necessary to ensure the safety of patients, as the number of patients in asylums increased, the use of physical restraints became more about controlling overcrowding rather than keeping patients safe. With the increase in the number of people considered mentally ill and the subsequent construction of more asylums, a new field of medicine called psychiatry emerged, with many medical professionals working to find cures for mental illness.

During the early twentieth century, hydrotherapy became a widely used medical treatment in many institutions. Hydrotherapy involves using water, ice, or steam, either internally or externally, to promote health or treat various diseases. The treatment varies in temperature, pressure, duration, and application site.

Medical professionals were eager to find a cure and had some interesting theories about how to do so. Some believed that infections were the leading cause of madness, and therefore suggested removing body parts such as teeth, tonsils, gallbladder, thyroids, appendix, and part of the colon to get rid of the infection. They believed any infected body part needed to be removed to cure the disease.

Another method was to inject a patient with malaria-infected blood to induce high fevers to fight off any signs of mental illness they might have. This method started to fade out when insulin shock therapy became widespread. This new treatment would inject a patient with high levels of insulin, resulting in convulsions and then a coma. They would do this for months until the patients were "cured" of their illness. Also, metrazol shock therapy was used to produce seizure-like convulsions in patients, therefore shocking their brains out of mental illness. What would replace metrazol would be electroconvulsive shock therapy. This carried a lesser risk of fracture in a patient. This method showed improvements in a patient's mood in just one treatment and was considered groundbreaking.

During the period when ECT was on the rise as a treatment for mental illness, another method called lobotomy also gained popularity. This surgical procedure involved severing the connections between the frontal lobe and other brain parts. Although doctors knew that patients undergoing this procedure would likely experience a loss of personality, motor skills, and memories, they believed that it was necessary to cure mental illness. Lobotomies were also thought to be effective at stopping unruly behavior and restoring sanity. However, the devastating side effects of this procedure ultimately led to its decline in popularity.

In the 1950s, the American mental health system introduced thorazine, which was the first antipsychotic drug. This new treatment was a significant milestone in treating patients with schizophrenia and other disorders. It was not limited to treating schizophrenia; it also could calm down even the most unruly patients with bad behavior, providing safety and peace for both staff and other patients. This treatment became known as a "chemical restraint," which was considered more humane than the physical restraints used in the past. Thorazine paved the way for other new types of antipsychotic drugs like lithium and prozac, which are popularly used today.

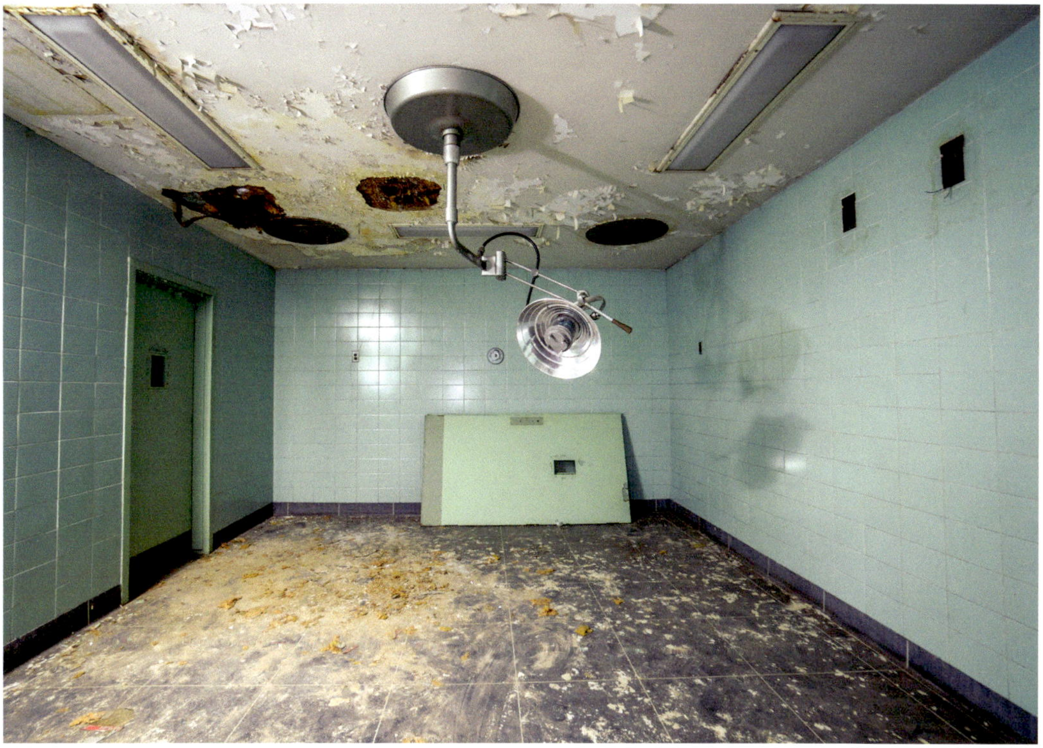

Around the 1960s, surgeries were typically performed in a controlled environment with artificial lighting and air.

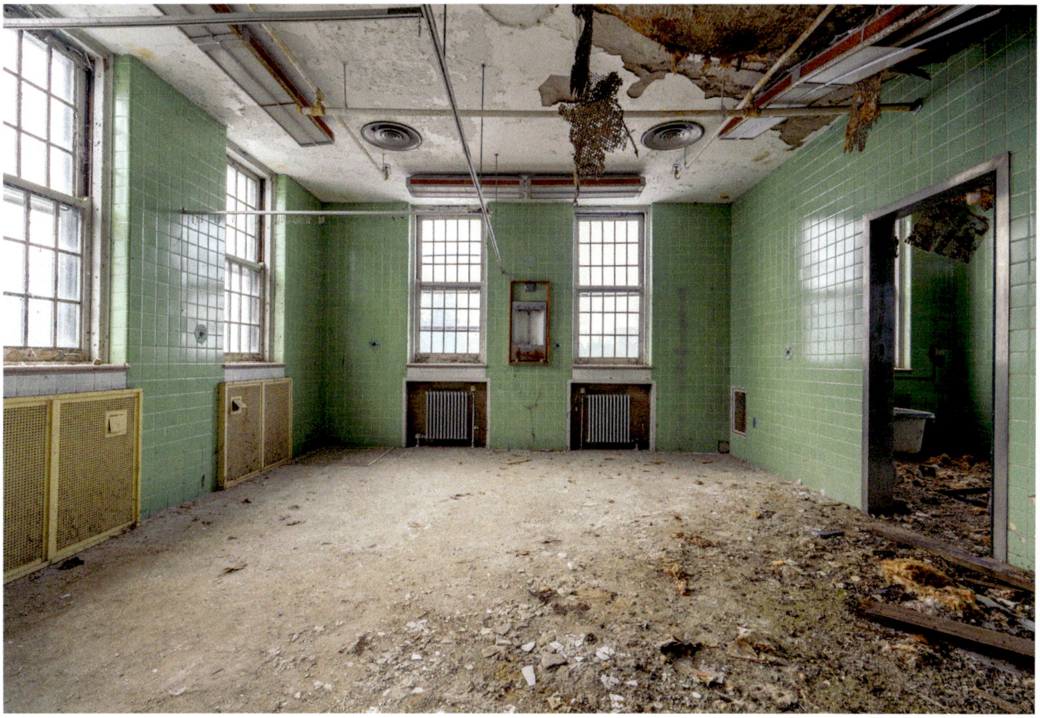

An infirmary ward was important to take care of sick or injured patients.

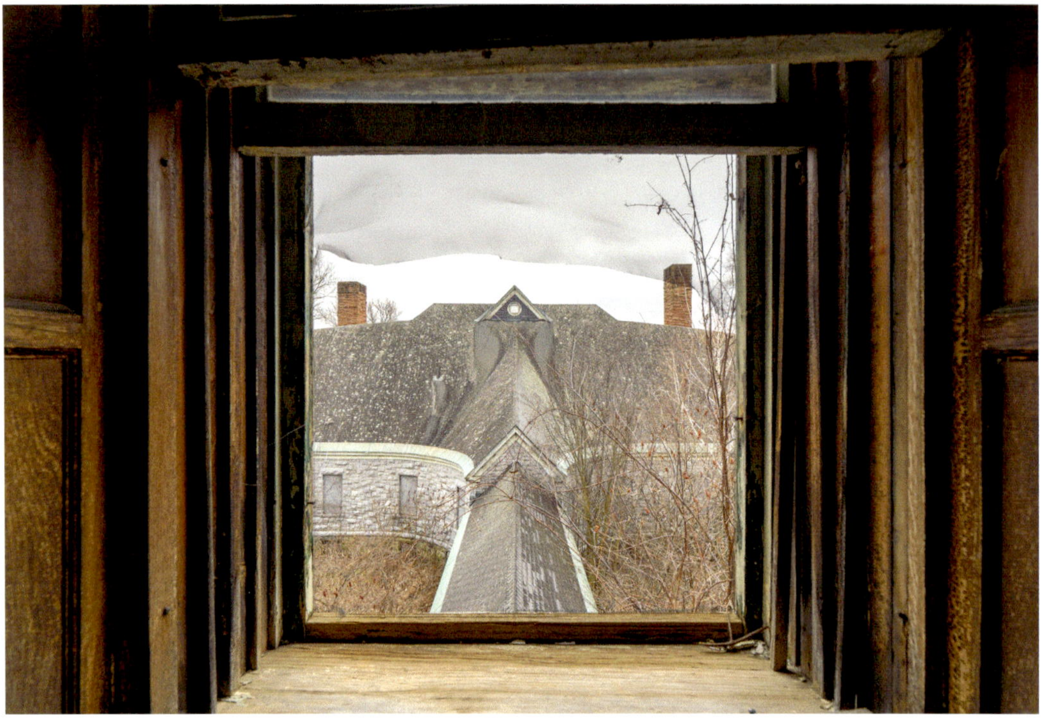

The rear window of the administration building overlooks a connector to the medical library and the hospital.

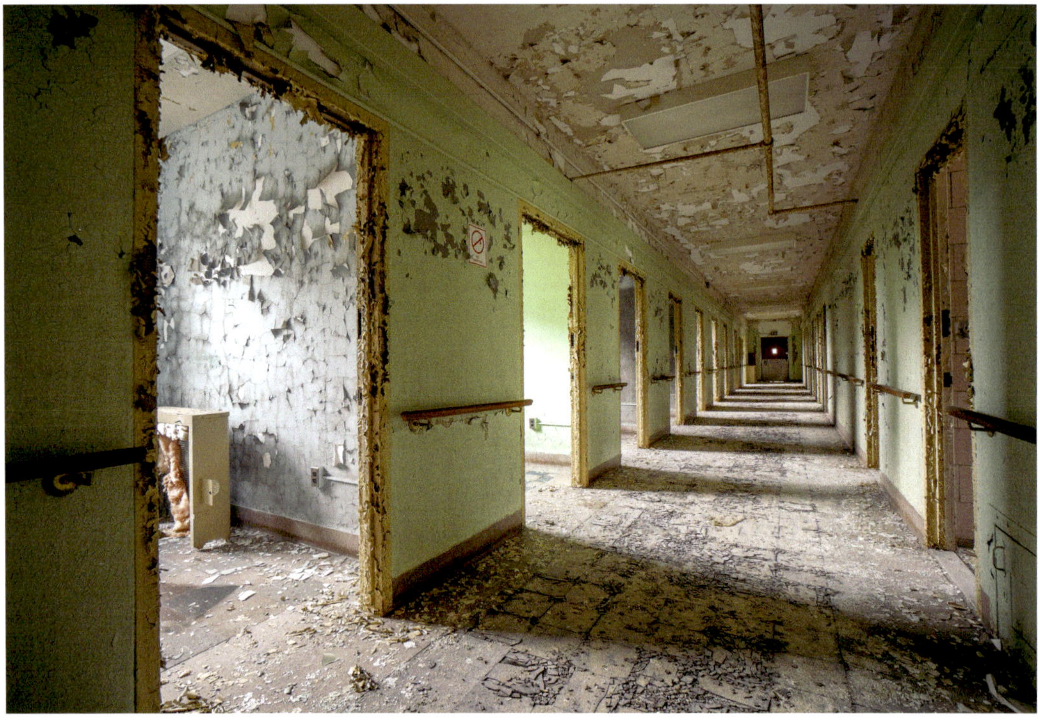

This medical building hallway had many rooms for patients waiting for surgeries, lab work, or checkups.

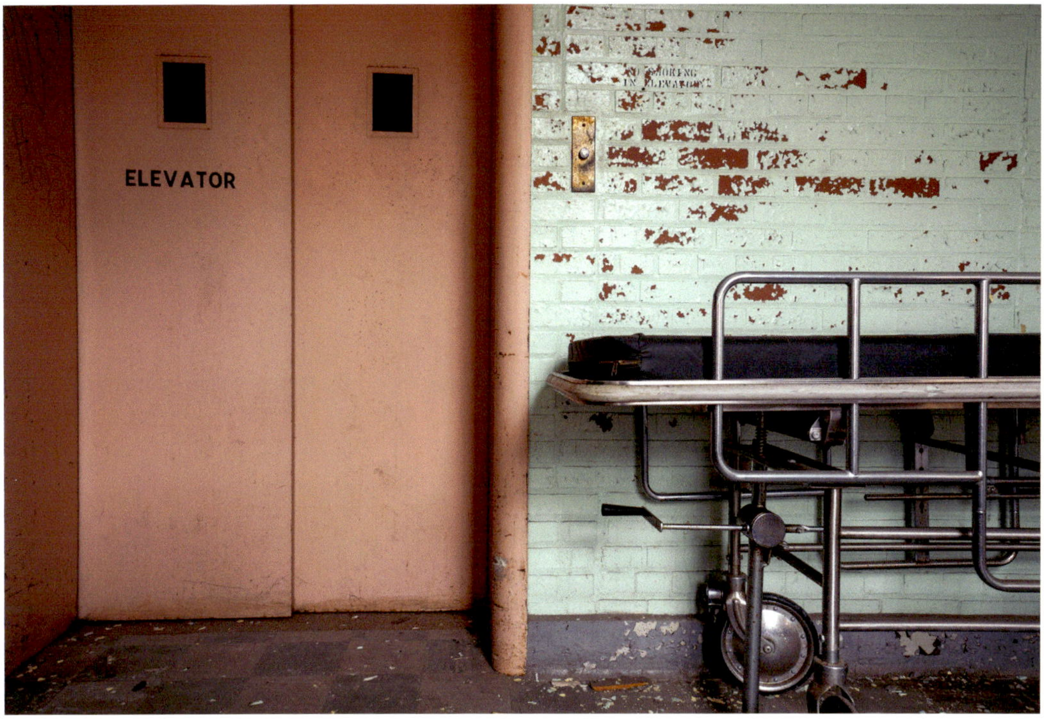

Elevators and gurneys were essential and the most efficient way to move sick patients between different wards and floors.

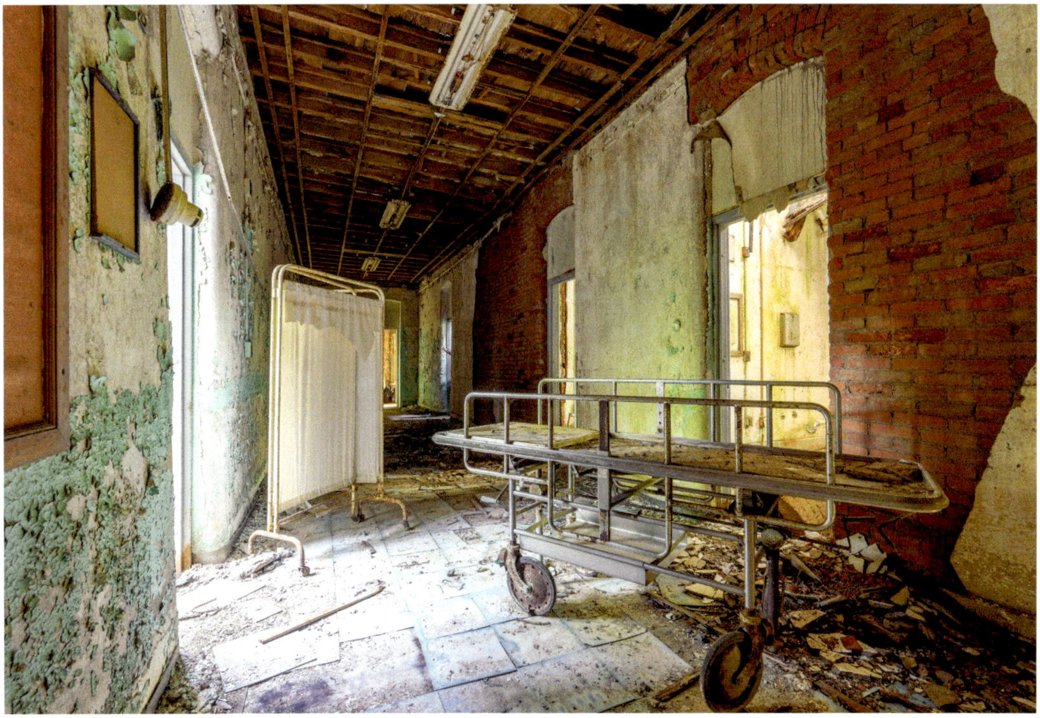

Hospital gurneys typically feature side rails and a manually adjustable backrest for enhanced safety and comfort.

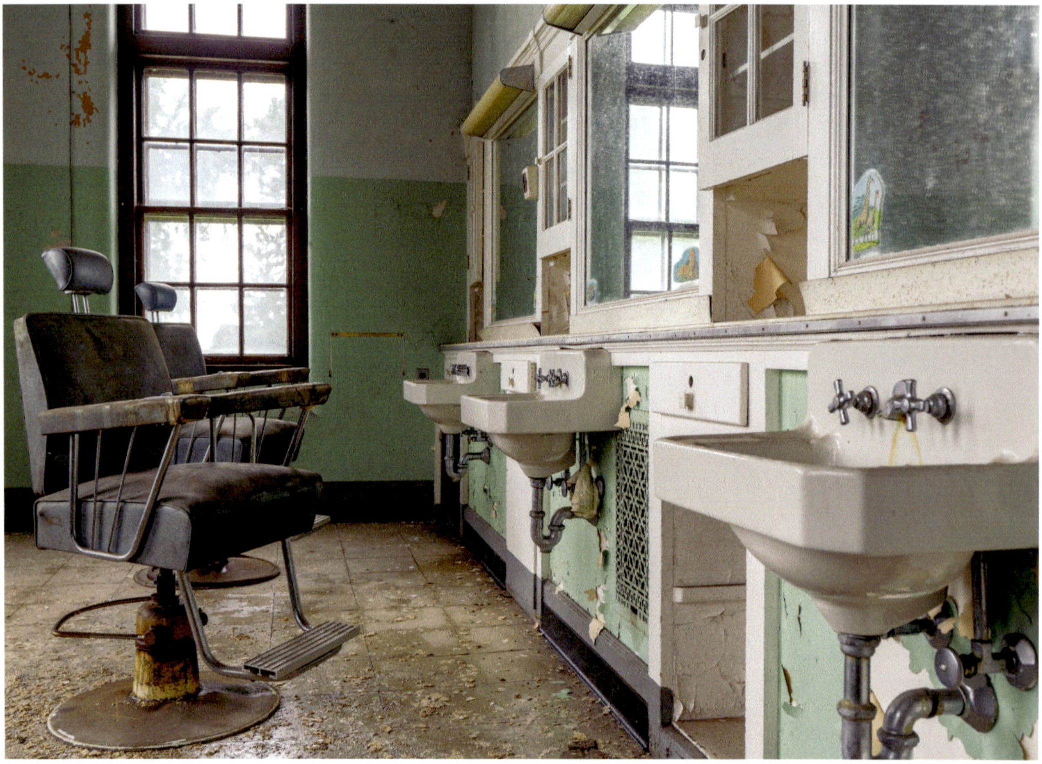

This institution established the first beauty salon for patients, which was later adopted into other asylums.

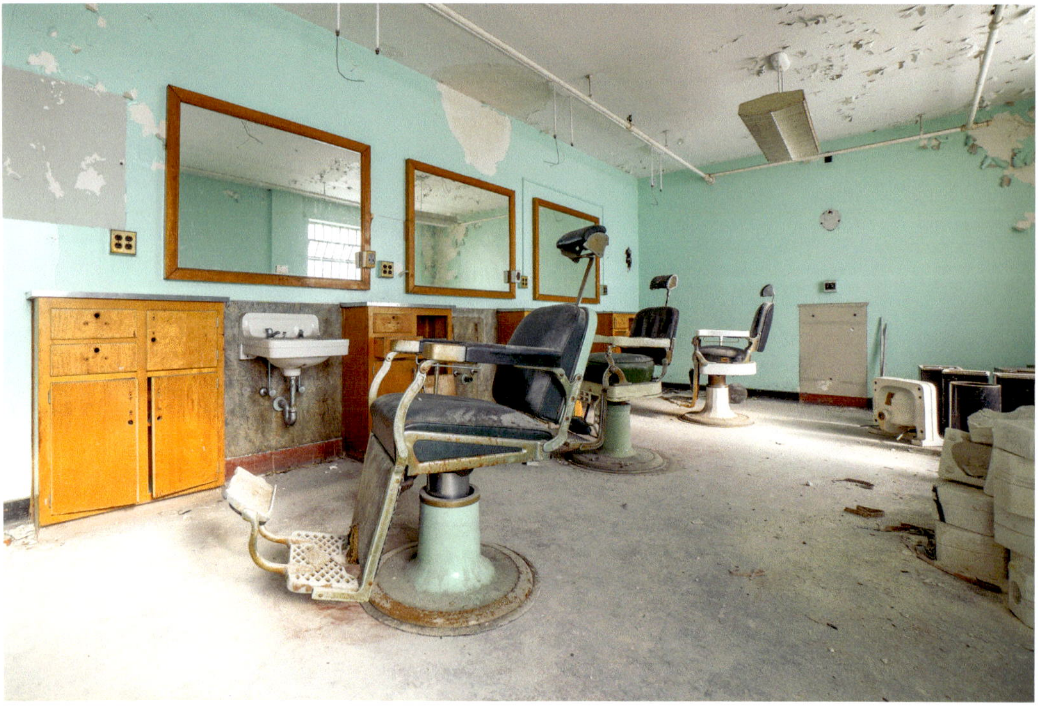

Looking nice and feeling good would help boost a patient's mental health.

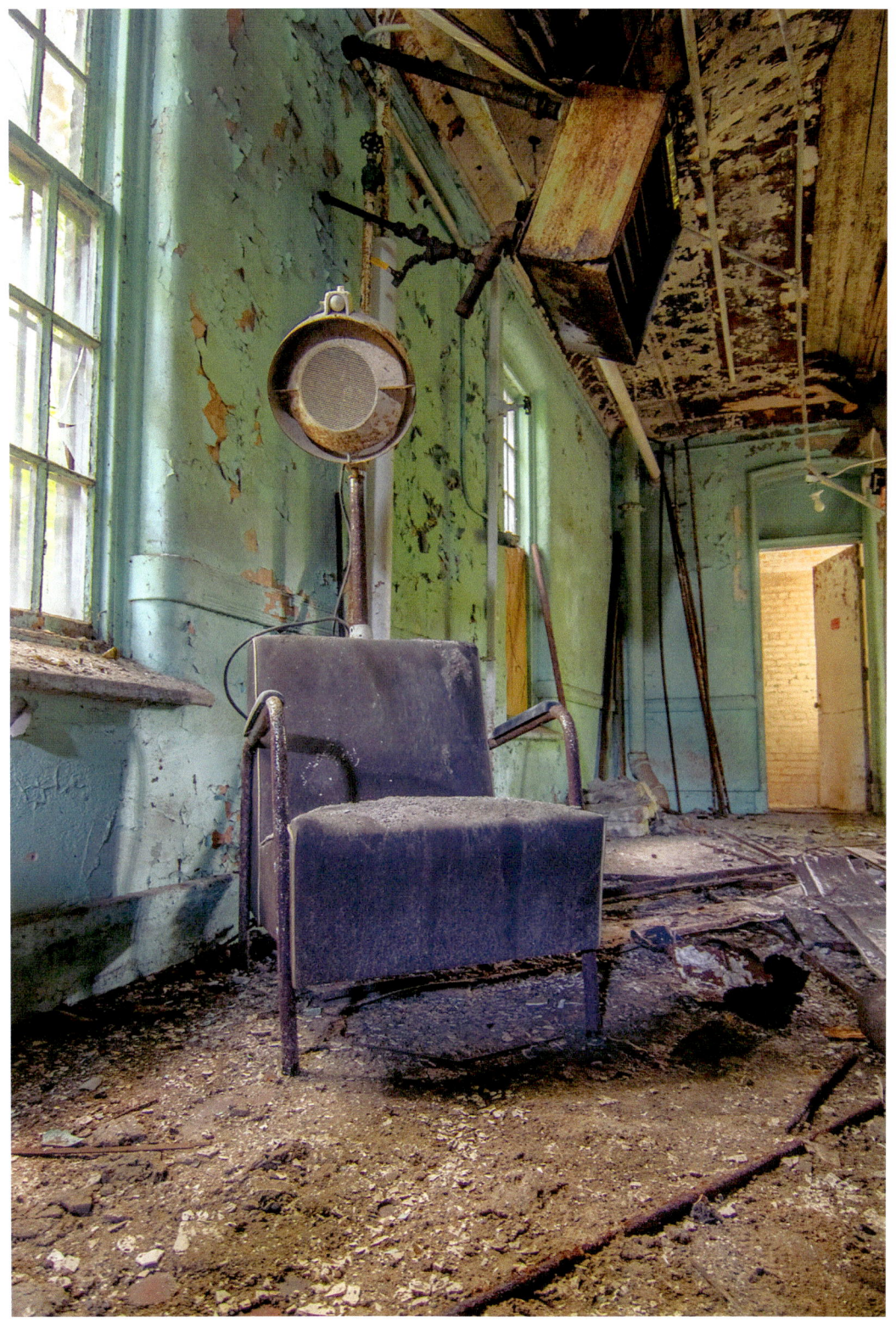
Beauty parlors for patients were popular and helped build social skills.

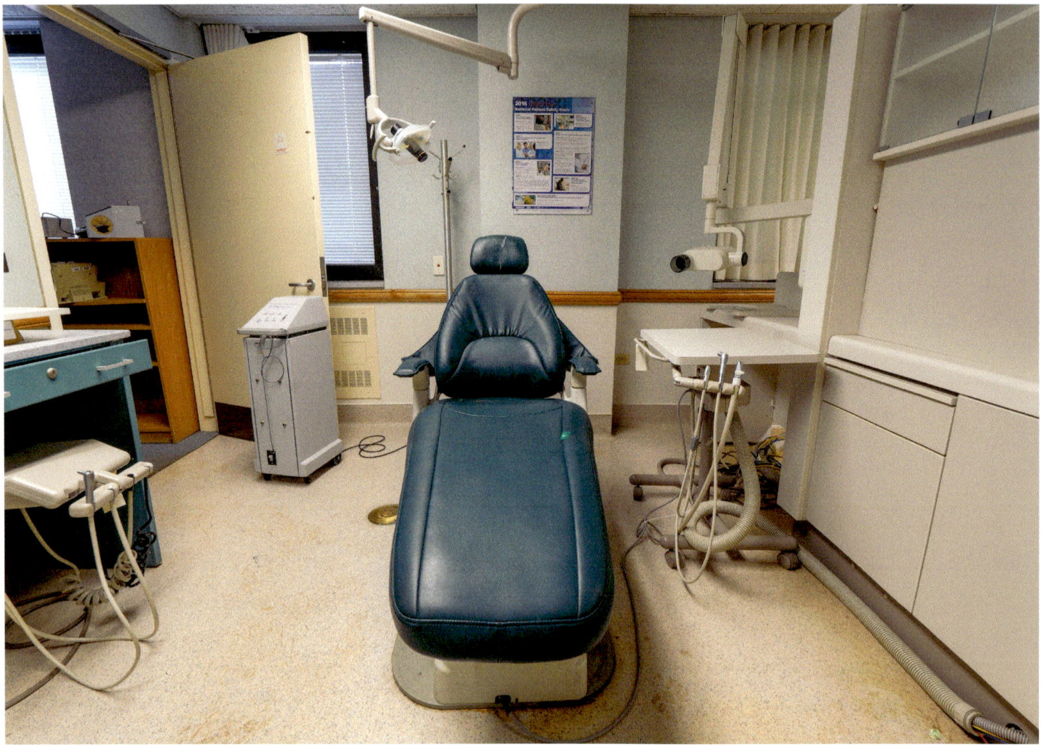

Dental work was also important for recognizing and identifying diseases also along with routine cleanings.

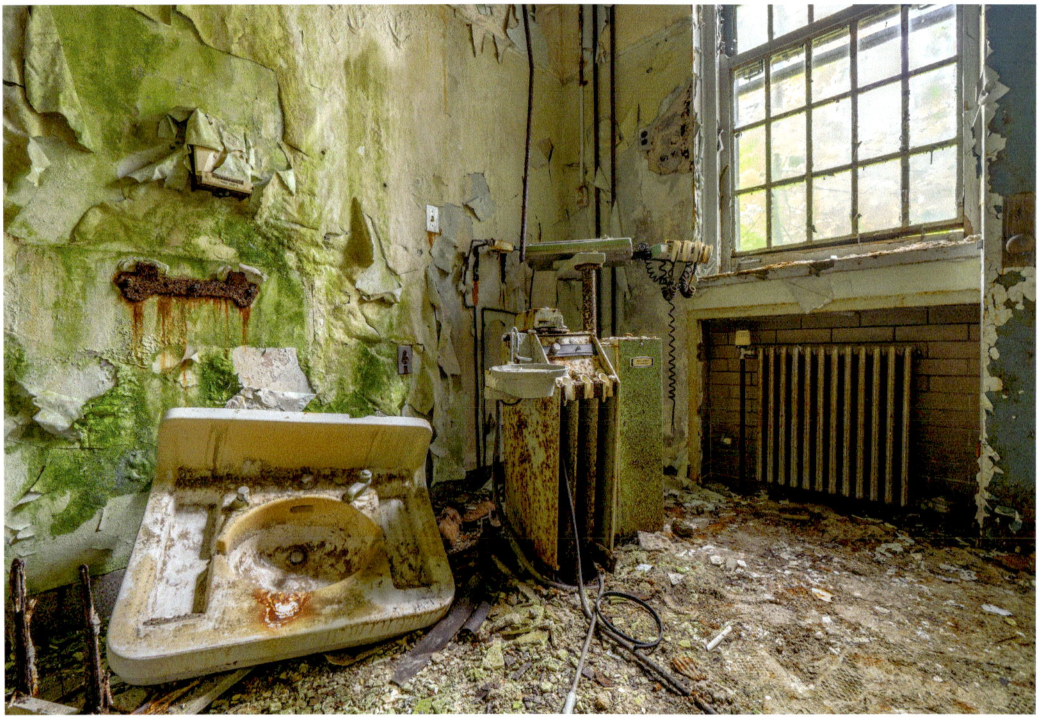

Some doctors thought that pulling infected teeth would "cure" mental illness.

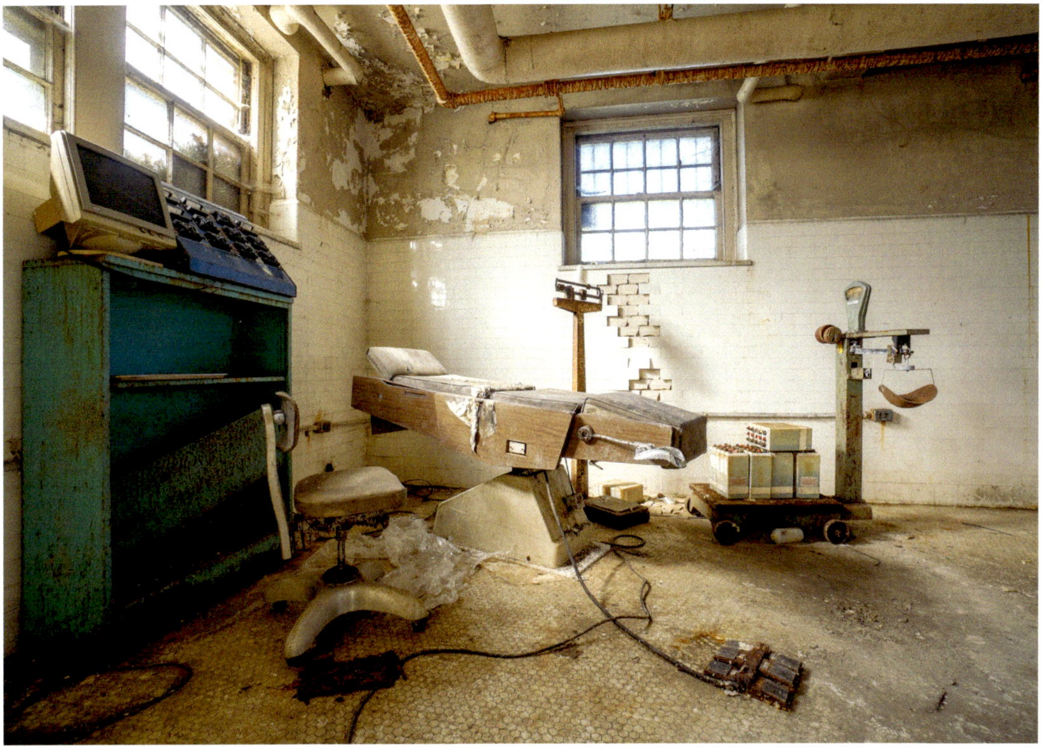

Medical equipment like gynecological examination chairs and scales were necessary for recovery and treatment.

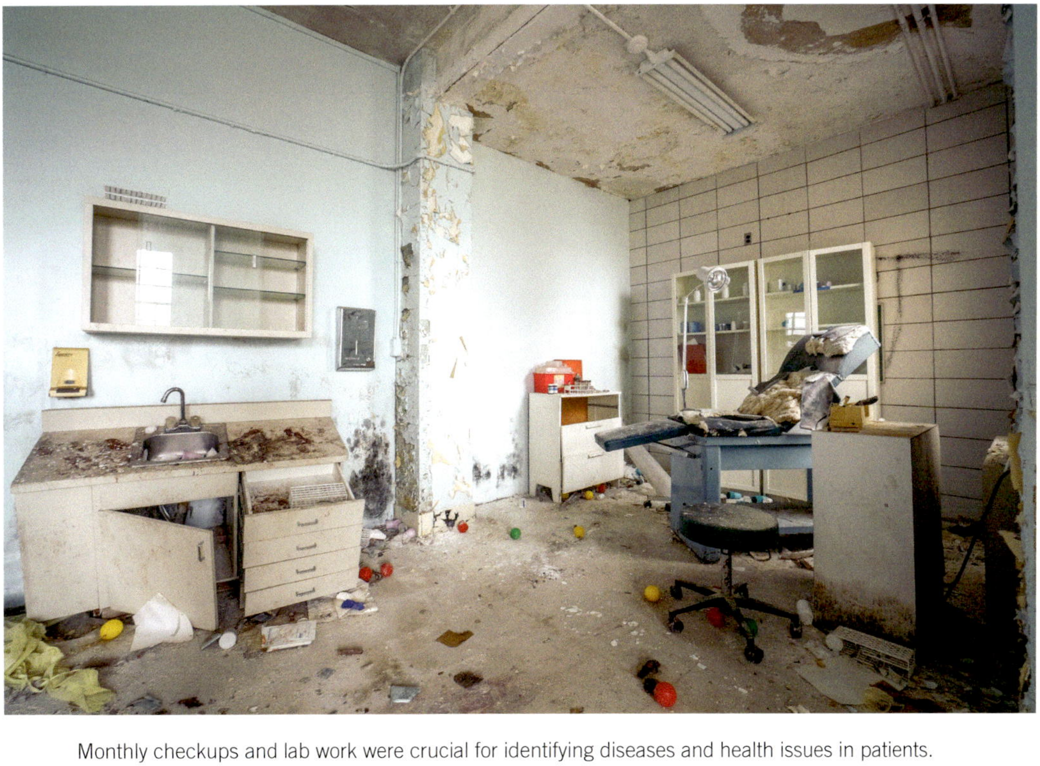

Monthly checkups and lab work were crucial for identifying diseases and health issues in patients.

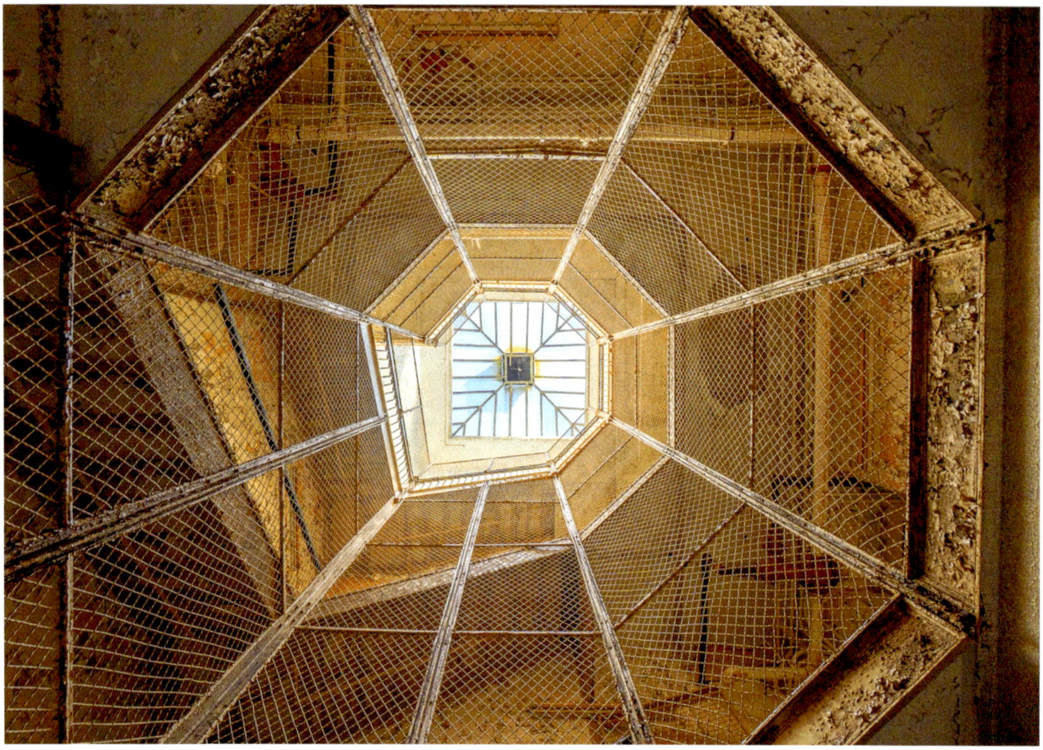

The risk of suicide is 50 times higher in psychiatric hospitals than in the general population.

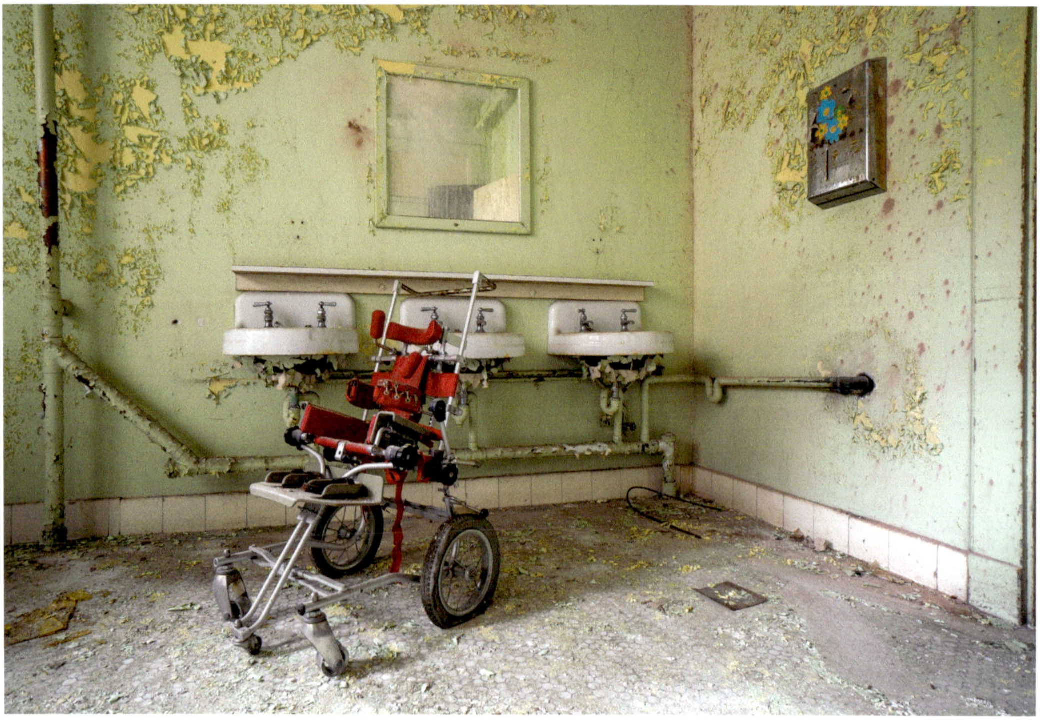

Special equipment was built or ordered for patients with certain medical conditions.

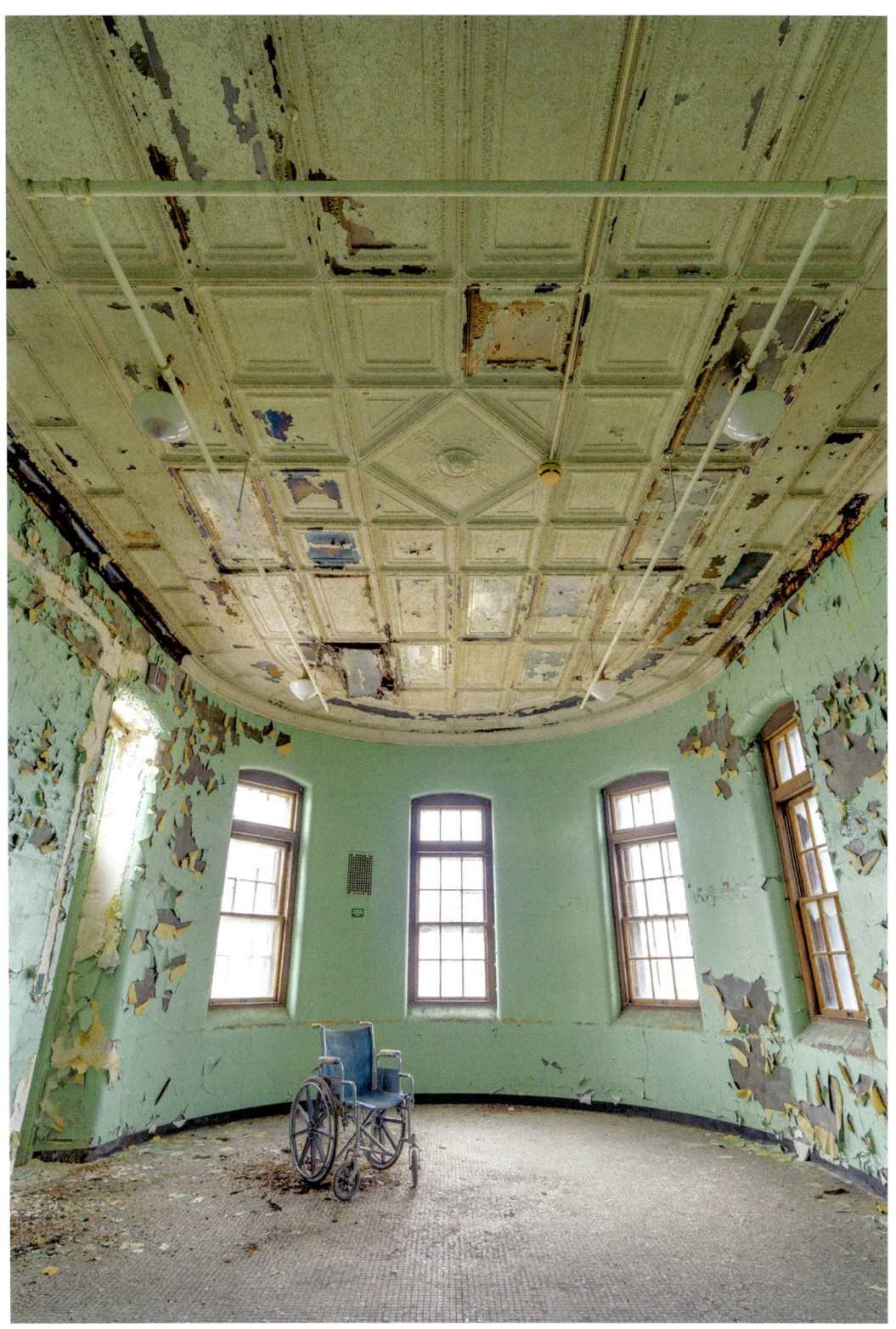
Wheelchairs were the most common and easily the most accessible for patient use.

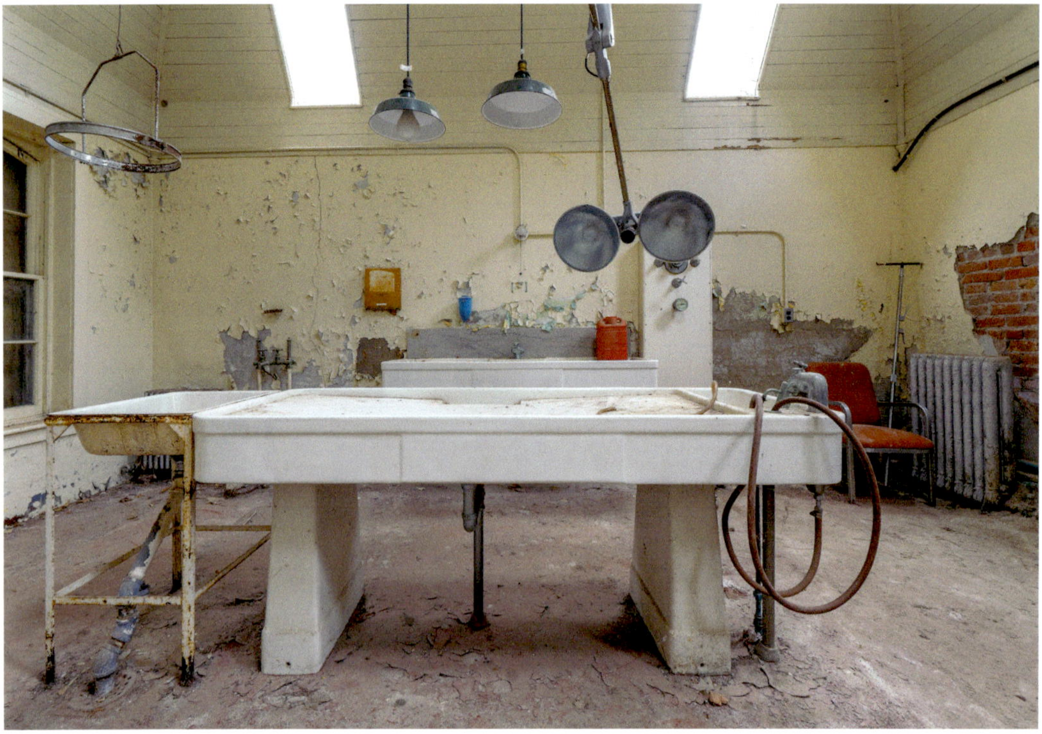

This autopsy table has processed over 6,000 patients in determining the cause of death.

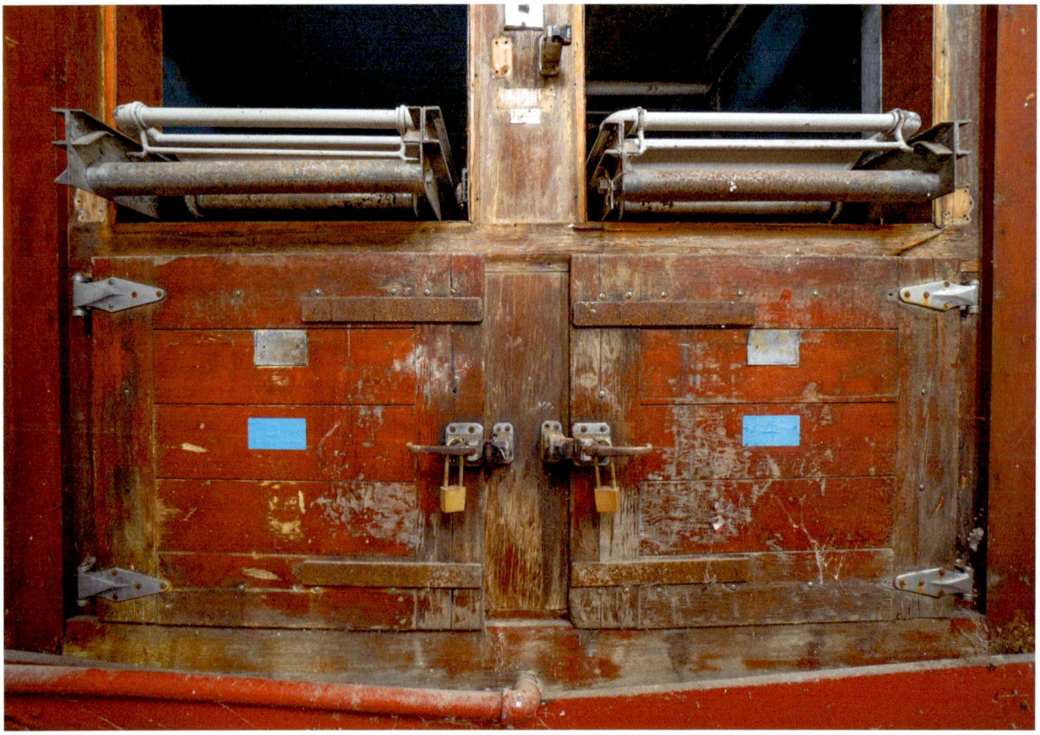

This 1870s morgue freezer would store deceased patients diagnosed as chronically insane.

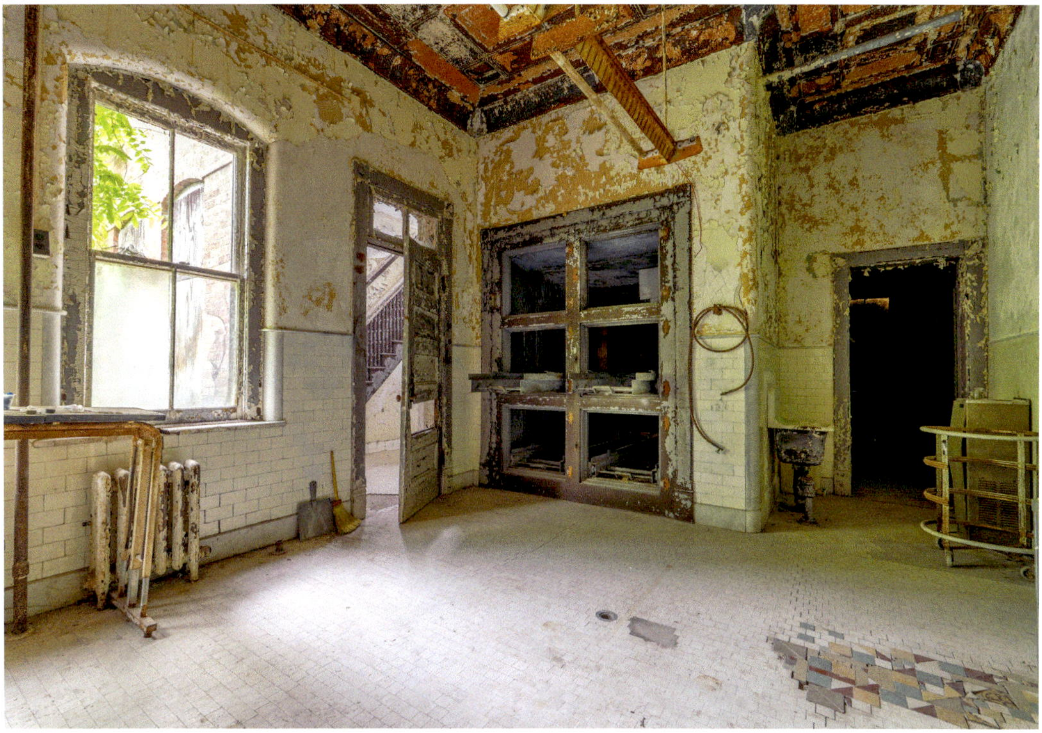

Asylum morgues were common and necessary for storing the deceased while waiting for an autopsy or burial.

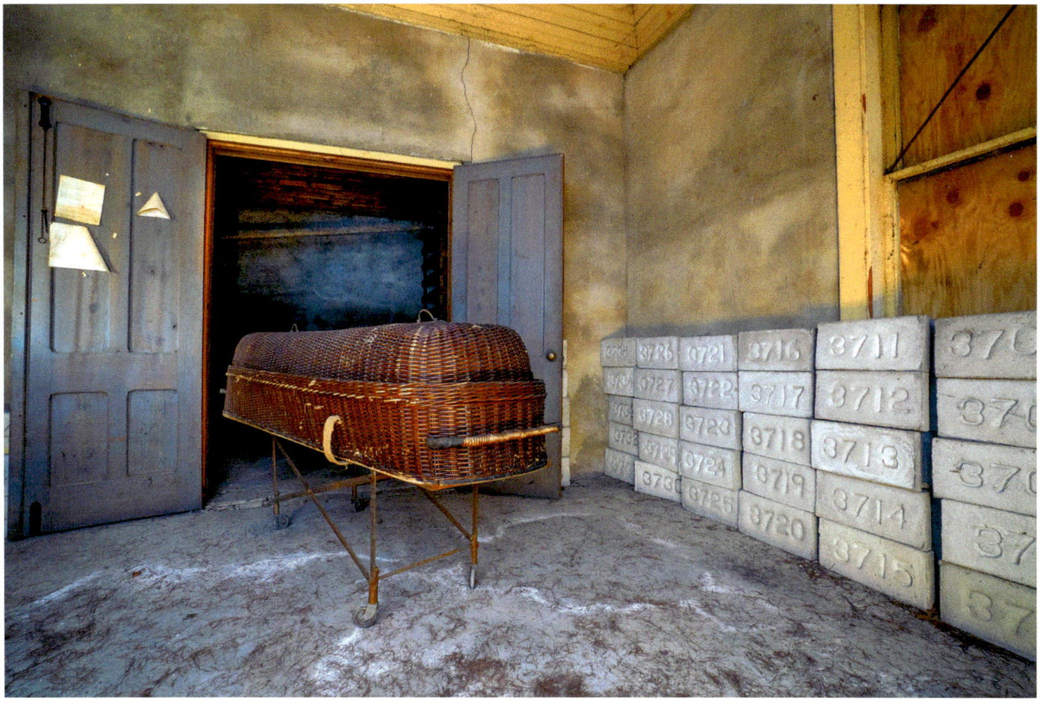

Whether a patient was buried in a casket or cremated, the resting place for the deceased was most likely on property or in a town plot.

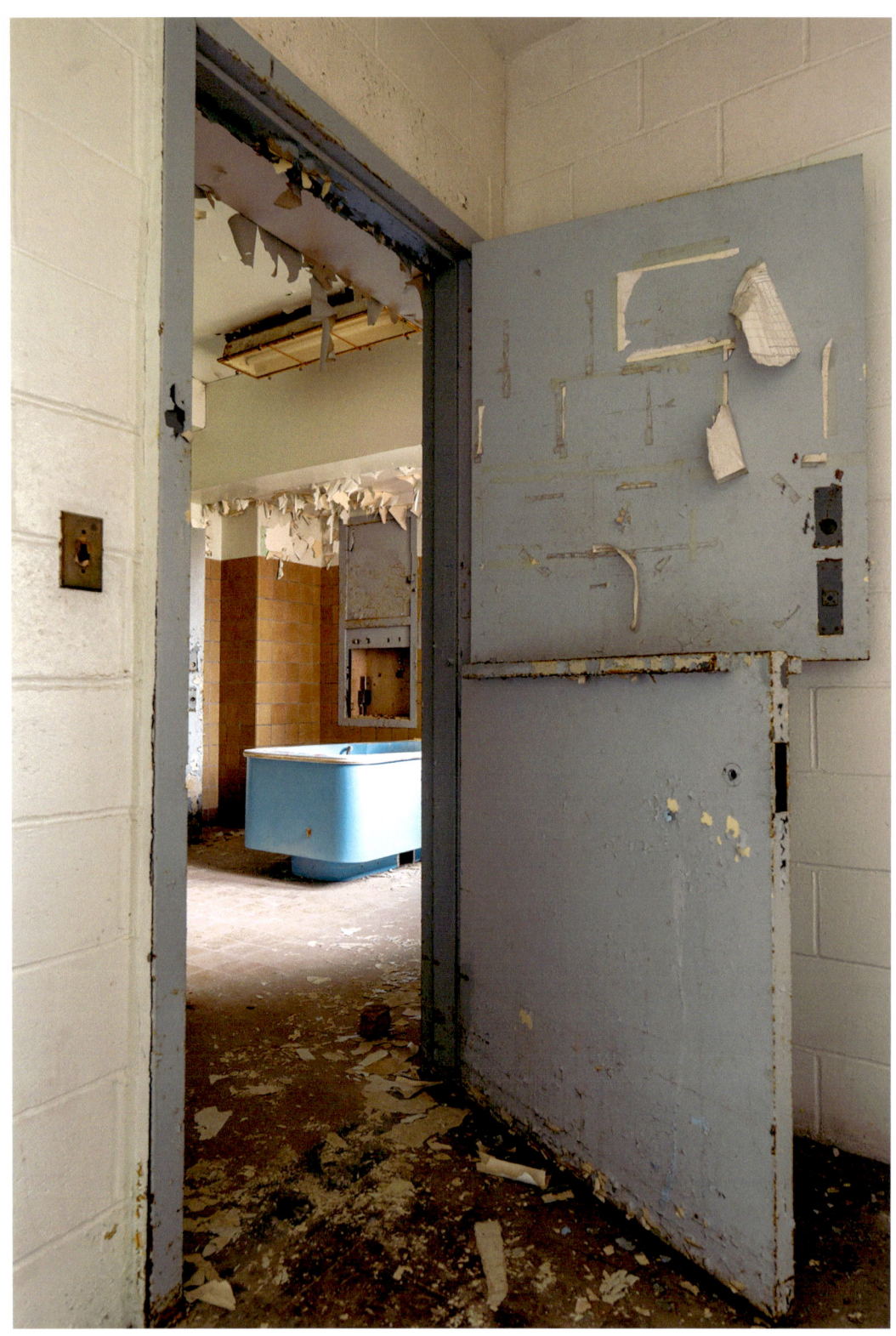
Hydrotherapy treatment became a popular treatment in the early 1930s.

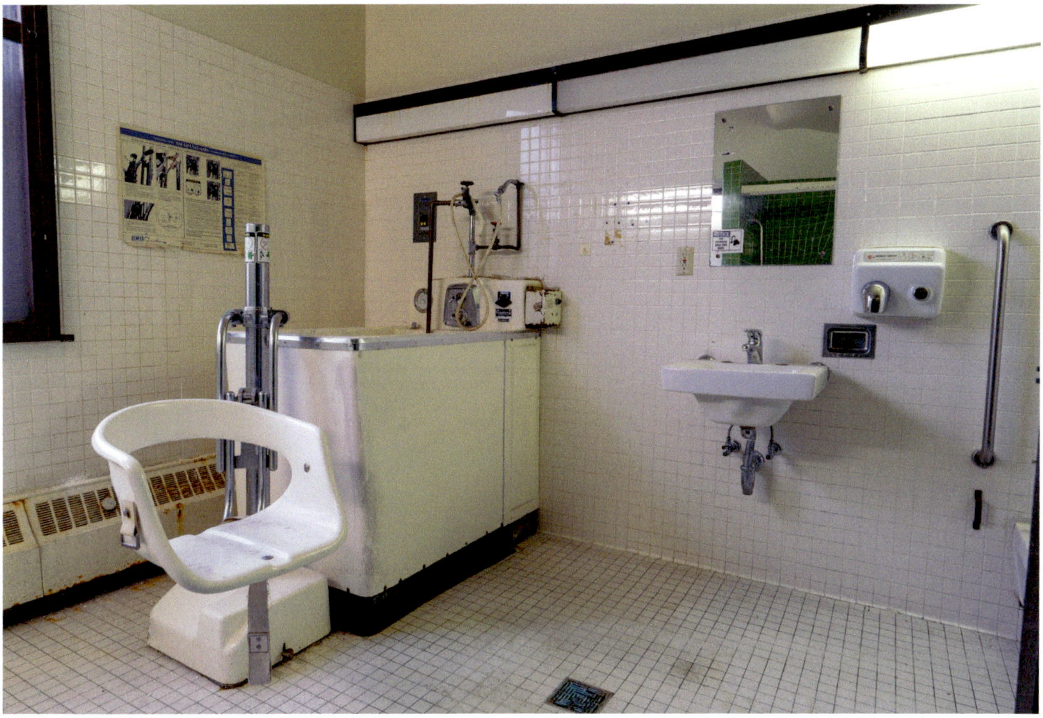

Exposing patients to warm baths for extended periods would help their mood and have a calming effect.

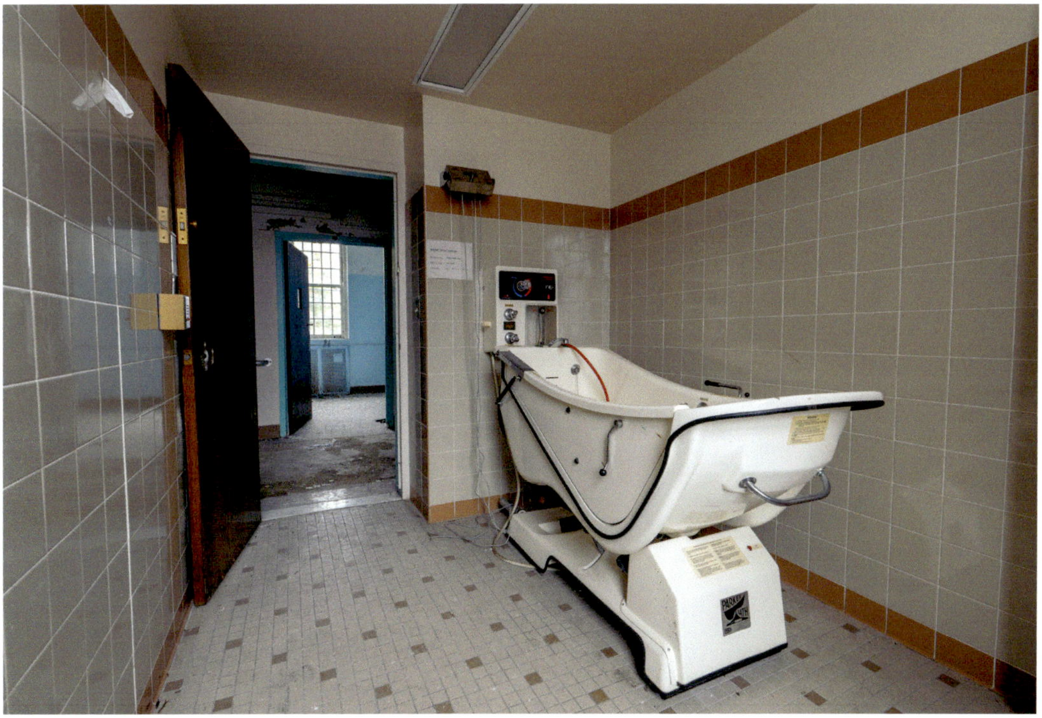

Although hydrotherapy is not as common as it once was, it is still used in mental hospitals to treat patients with PTSD.

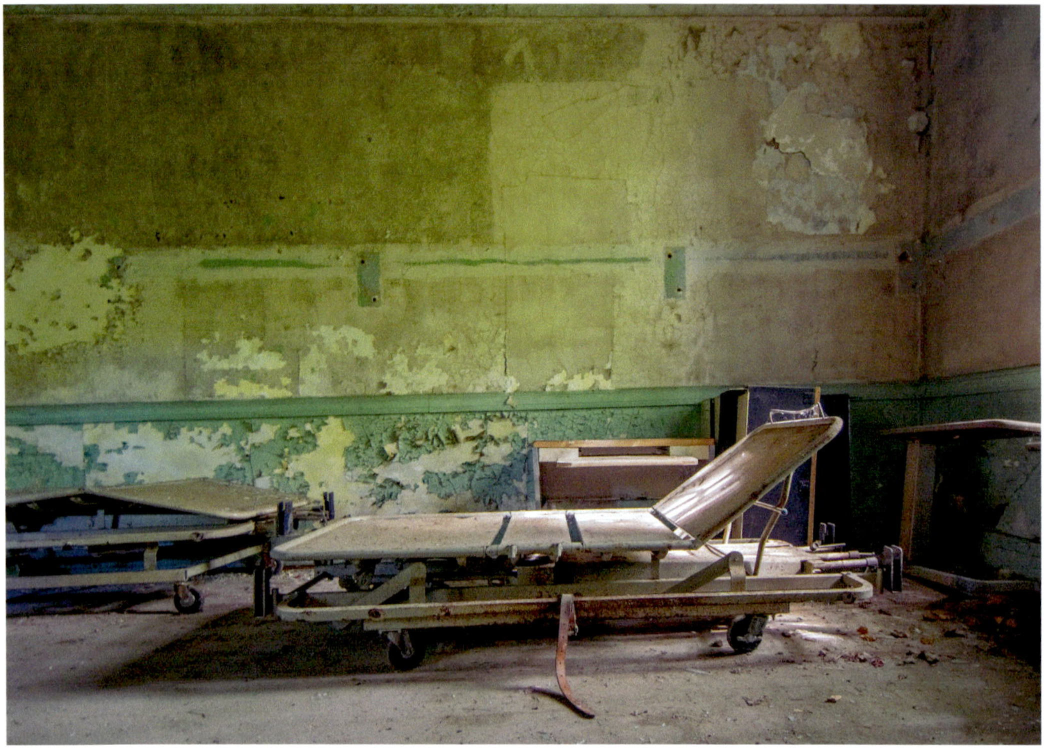

Medical restraints were used to protect patients from hurting themselves or staff.

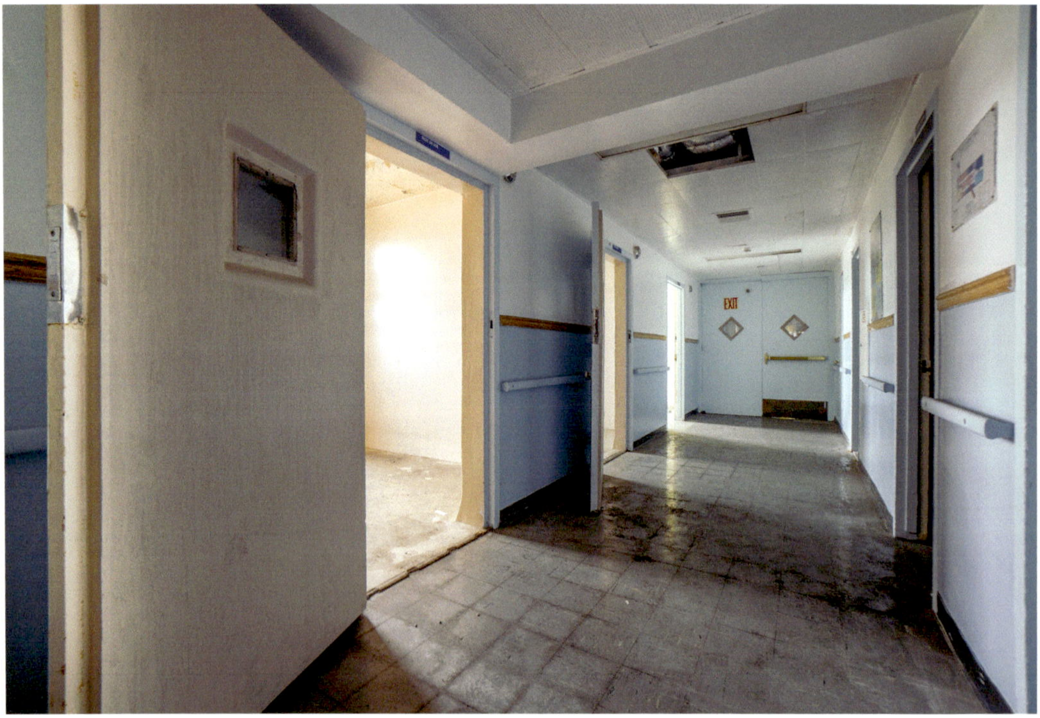

Isolation rooms became another method to help patients not hurt themselves with padded walls and floors.

# 3
# PLANS

Asylums and institutions were established in New York to provide stability and rehabilitation to patients with mental illness. Initially, the care for the mentally ill was mostly provided in private hospital wards or poorhouses in their respective counties. However, due to overcrowding and poor living conditions, two activists—Dorothea Dix and Dr. Thomas Kirkbride—stepped forward. Their efforts led to the introduction of moral treatment in mental health care in the United States. These two individuals played a significant role in creating the first generation of asylums in the country. In the state of New York, institutions were built mostly on three architectural styles: pre-1854, Kirkbride Plan, and Cottage Plan.

*Pre-1854 Plan*: Before 1854, there was no standard design or layout for institutions that housed mentally ill patients. These institutions were often groups of individual buildings located on the outskirts of busy cities. Most were separate buildings or wards in private hospitals or stand-alone structures.

*Kirkbride Plan*: The Kirkbride Plan was a groundbreaking development in the field of mental health care, designed around the idea of morally treating patients. The plan involved the architectural design of the institution, its location, ventilation, and natural lighting. Dr. Thomas Kirkbride, a visionary, designed the plan, which comprised a centrally located administrative building with wings attached on each side, one for male and the other for female patients. These wings were segregated based on the class of patients they housed, with the farthest wings accommodating the most "excited" patients, while the closer ones to the center were for the more stable patients. The center location of these buildings usually had chapels, auditoriums, libraries, and kitchens built directly onto the rear of the administration section. This was a testament to the innovative thinking of the time, as architecture and design were crucial for the treatment of the mentally ill.

*Cottage Plan*: The Cottage Plan type of institution was created in the 1890s to solve overcrowding issues. These campuses were designed more like a college campus, with multiple individual buildings that could be connected with above or underground tunnels. The aim was to separate patients with specific types of mental illnesses in different buildings, with an administrative building usually located at the center of the campus. Most cottage-style buildings did not have the castle-like design or intricate details of the popular Kirkbride Plan but instead had a more straightforward architectural style since it was not part of the treatment.

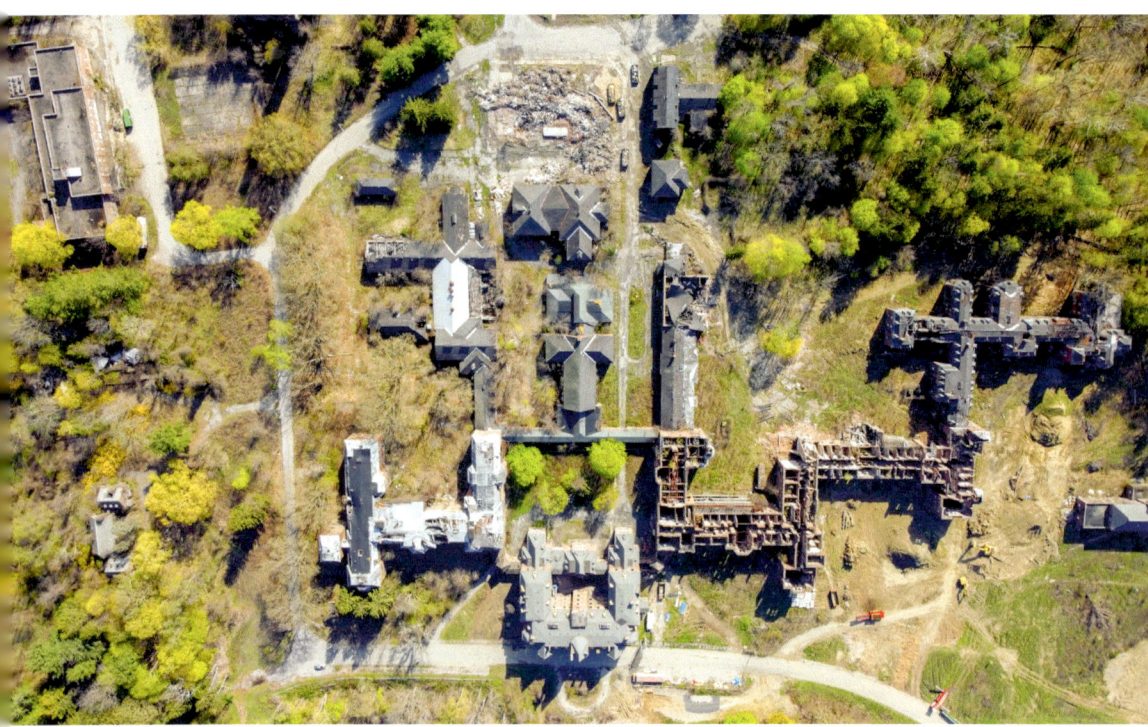

This aerial shot of a Kirkbride Plan asylum shows you the bird-like appearance of its massive wings and center.

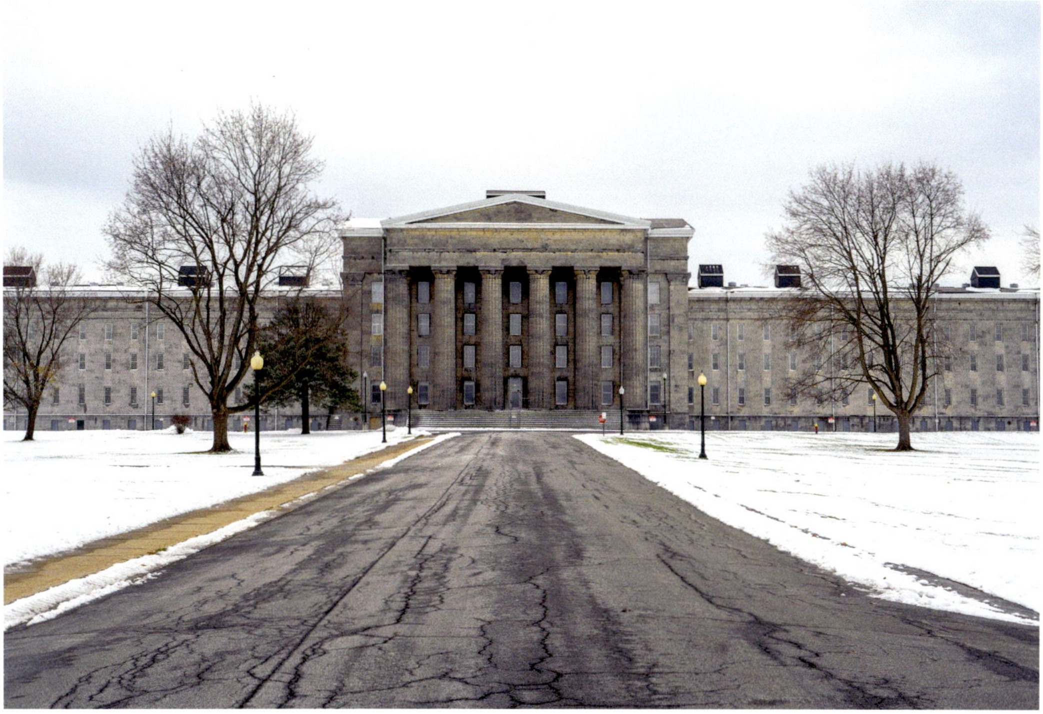

This asylum was the first state-run institution in New York specifically designed to care for the mentally ill.

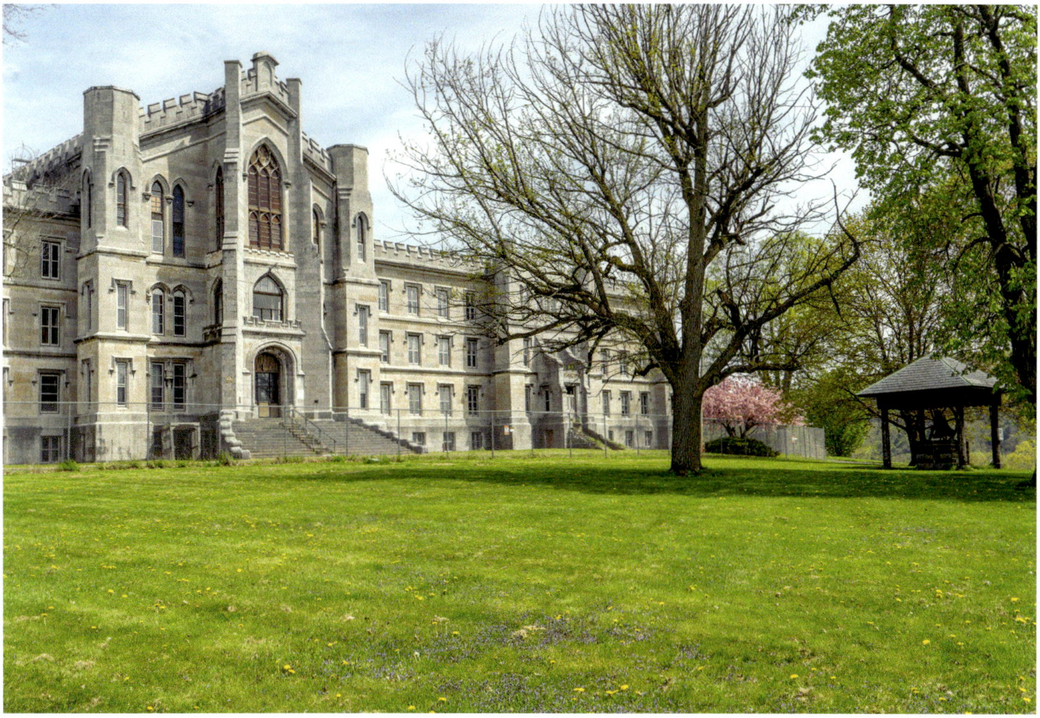

Originally designed as an institution for alcoholics, it only remained operational for 15 years before being repurposed as the second facility for the chronically insane in New York in 1881.

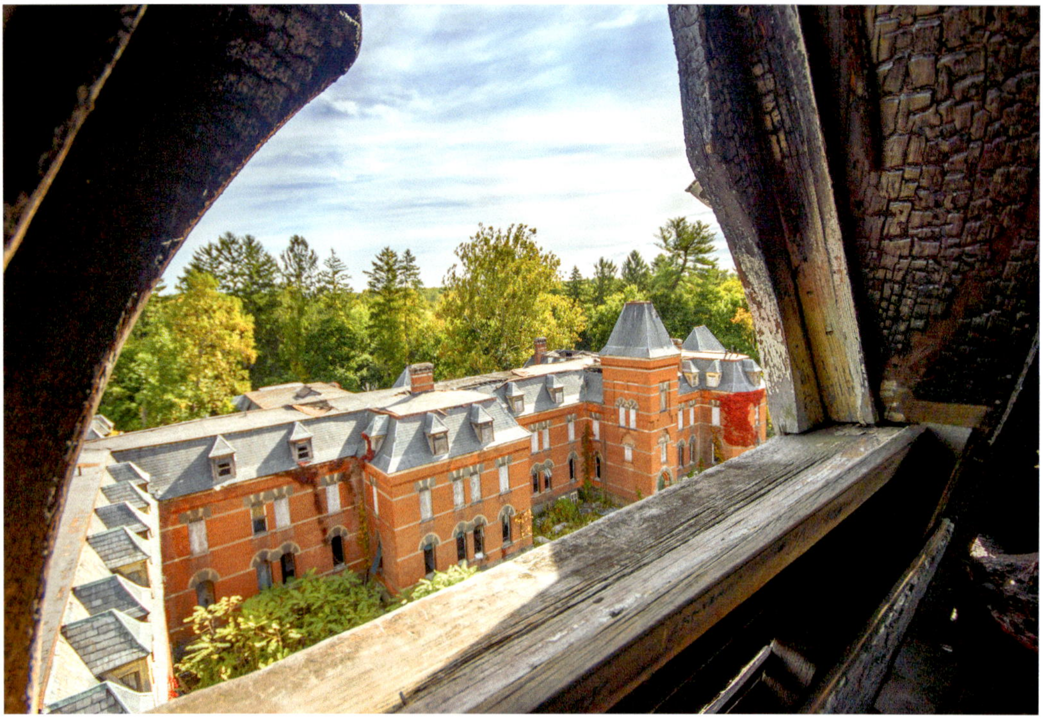

A distant view of the farthest ward in a Kirkbride building, which was reserved for the most "excited" patients.

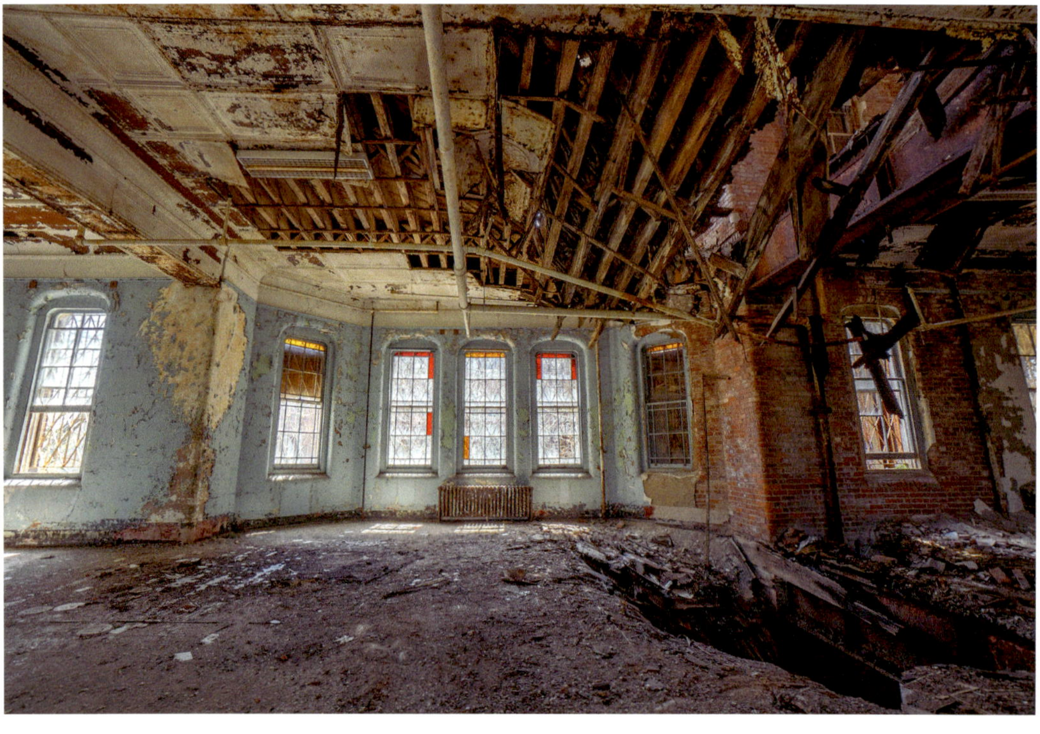

Open dormitories were an easy way to fit as many patient beds as possible when space was limited.

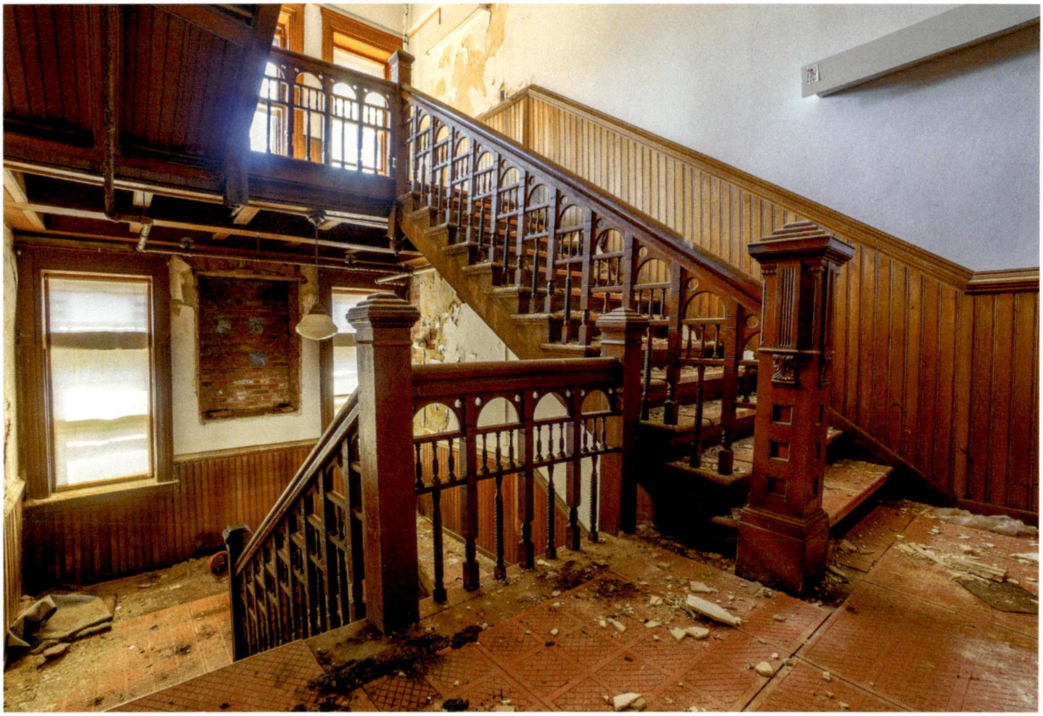

Many early asylum administration buildings featured intricately designed wooden staircases that highlighted the craftsmanship of their era.

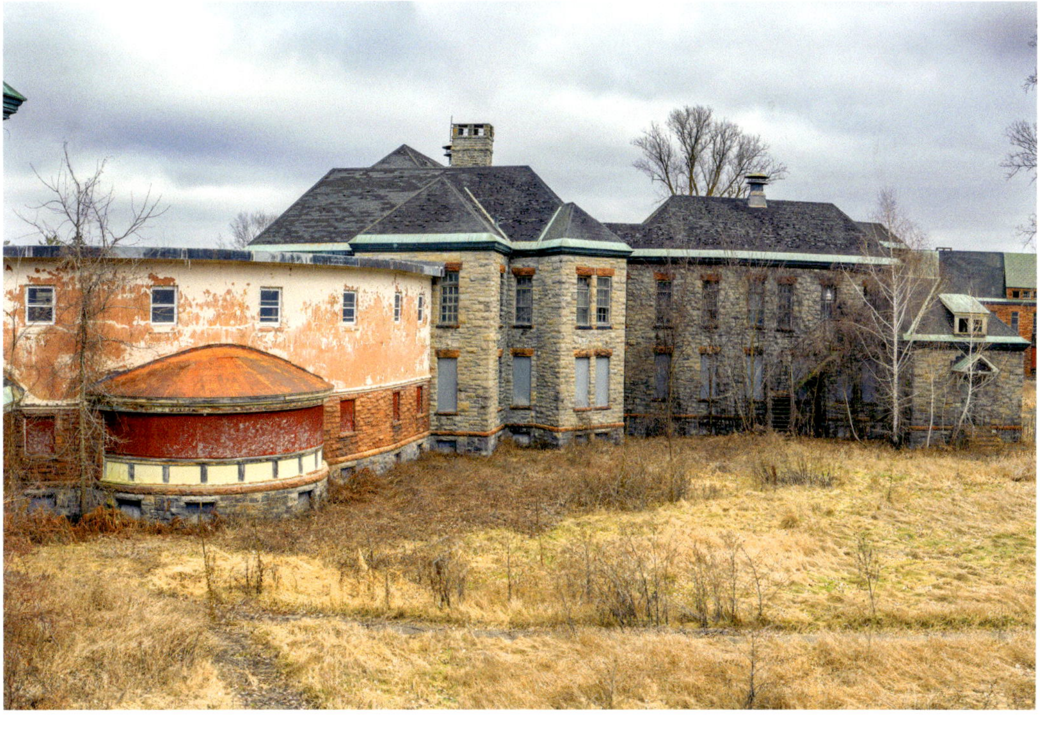

This cottage plan asylum takes on a Kirkbride plan feel with connecting breezeways and stonework.

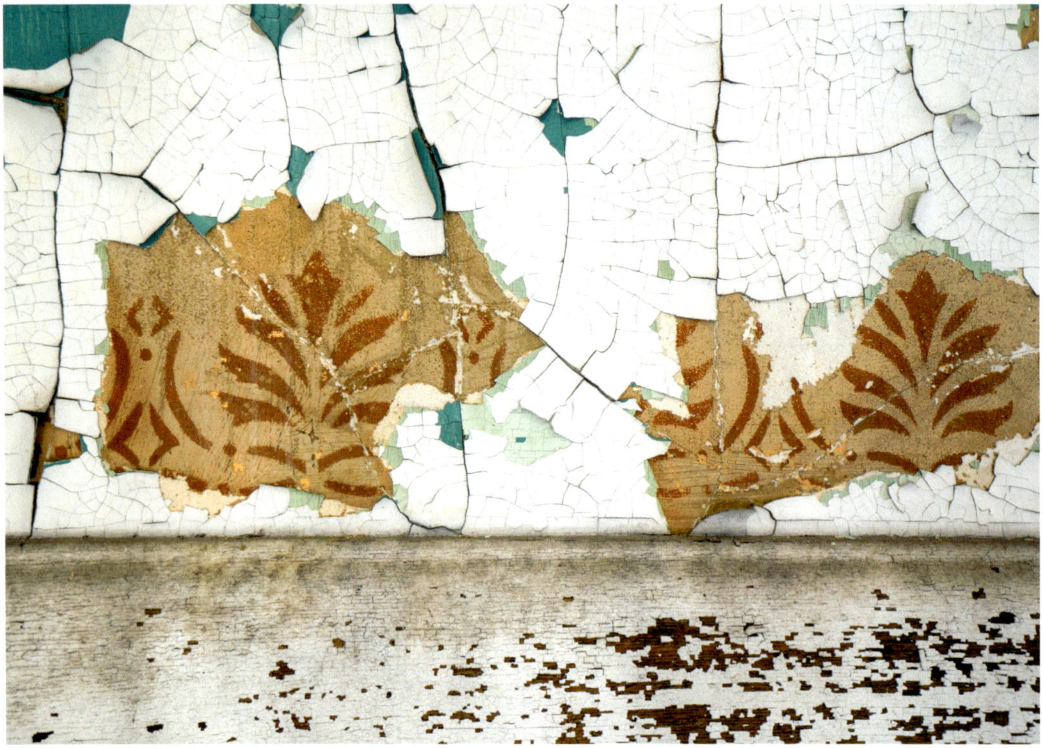

Decaying walls reveal hand-painted stencils that resurface beneath layers of paint.

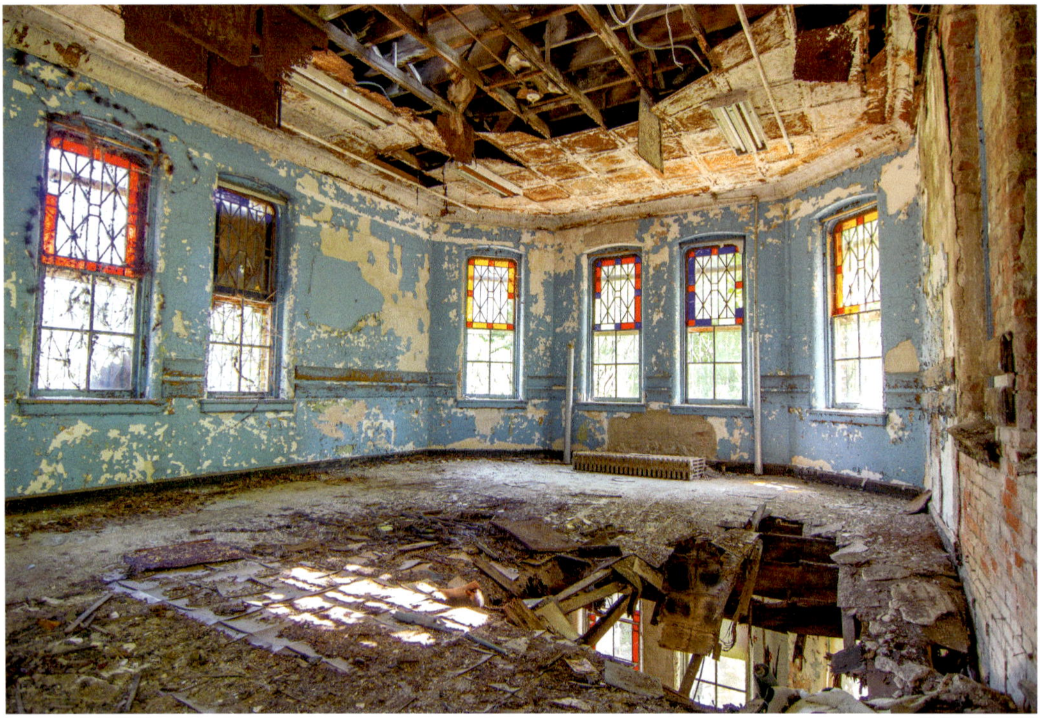

This dayroom is adorned with beautiful stained glass that adds to the serene atmosphere.

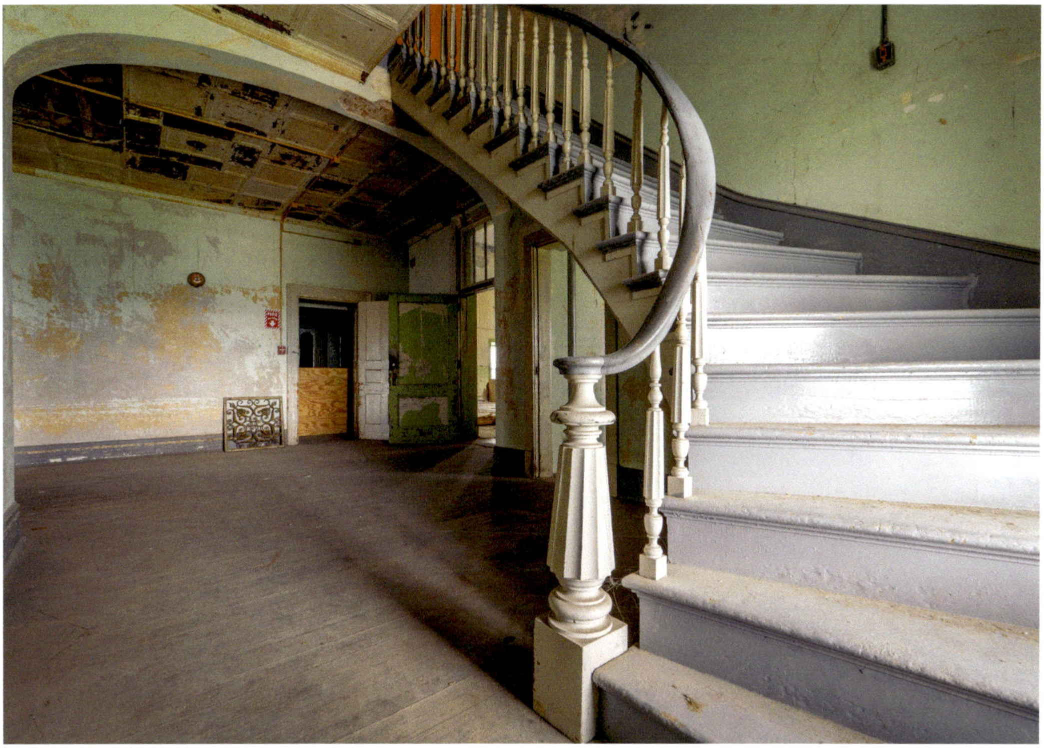

The popular Kirkbride plan is believed to have been influenced by New York's first institution's architecture.

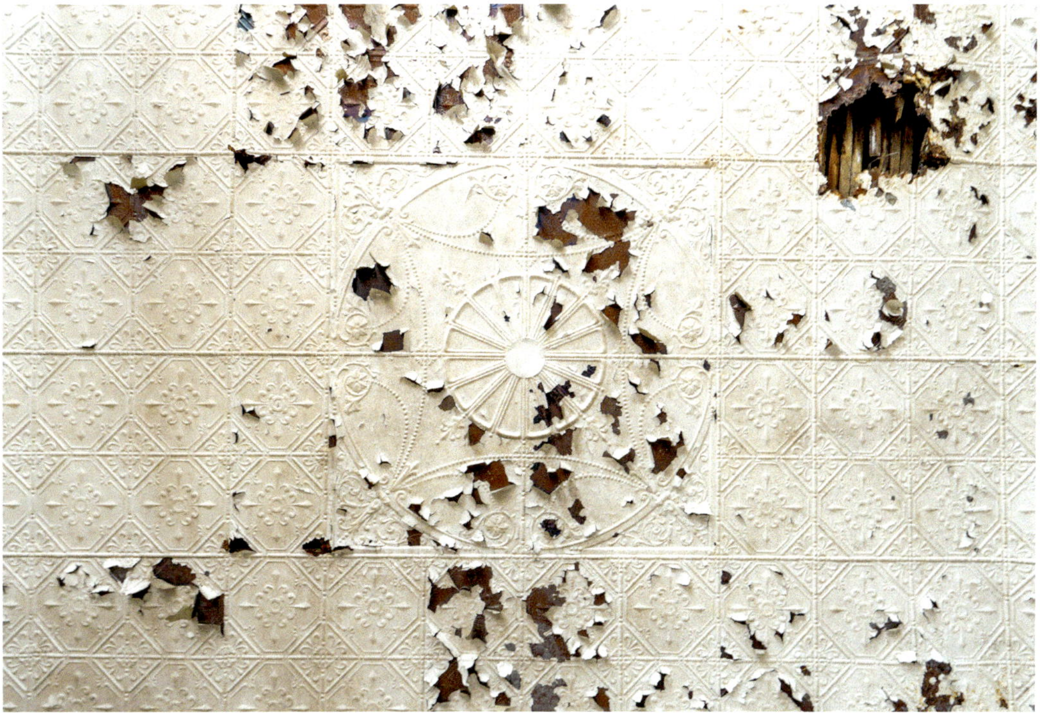

Tin ceilings promised to resist fire and improve hygiene and beauty at an affordable price.

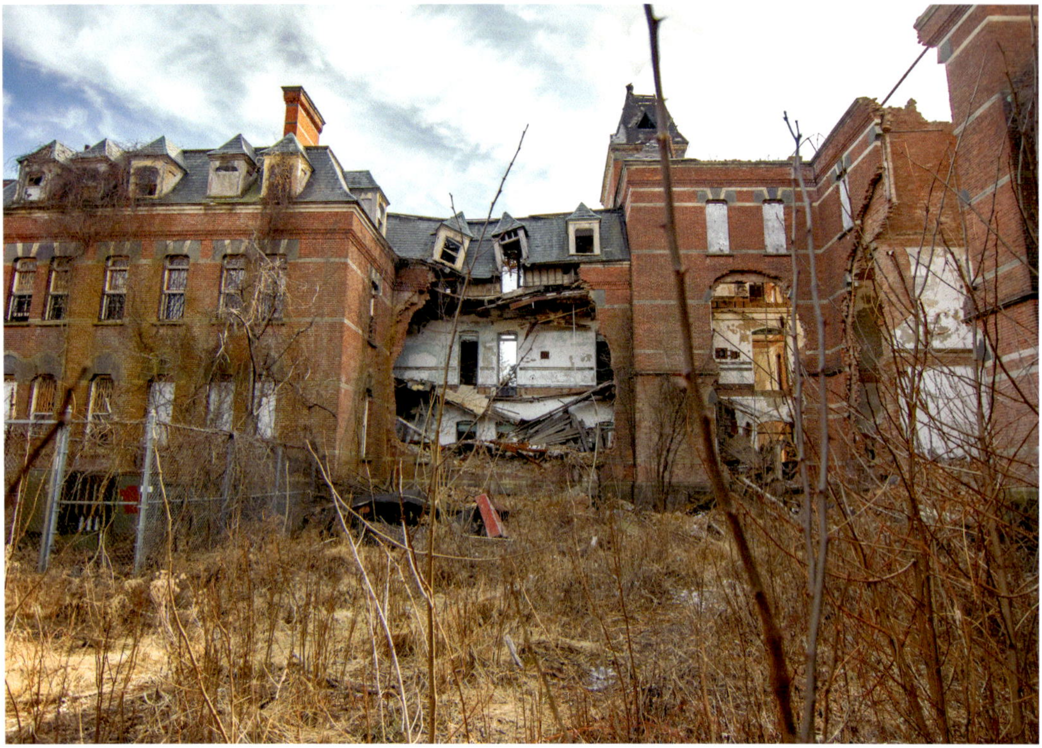

This gaping hole resulted from the fire department's attempt to contain a massive fire.

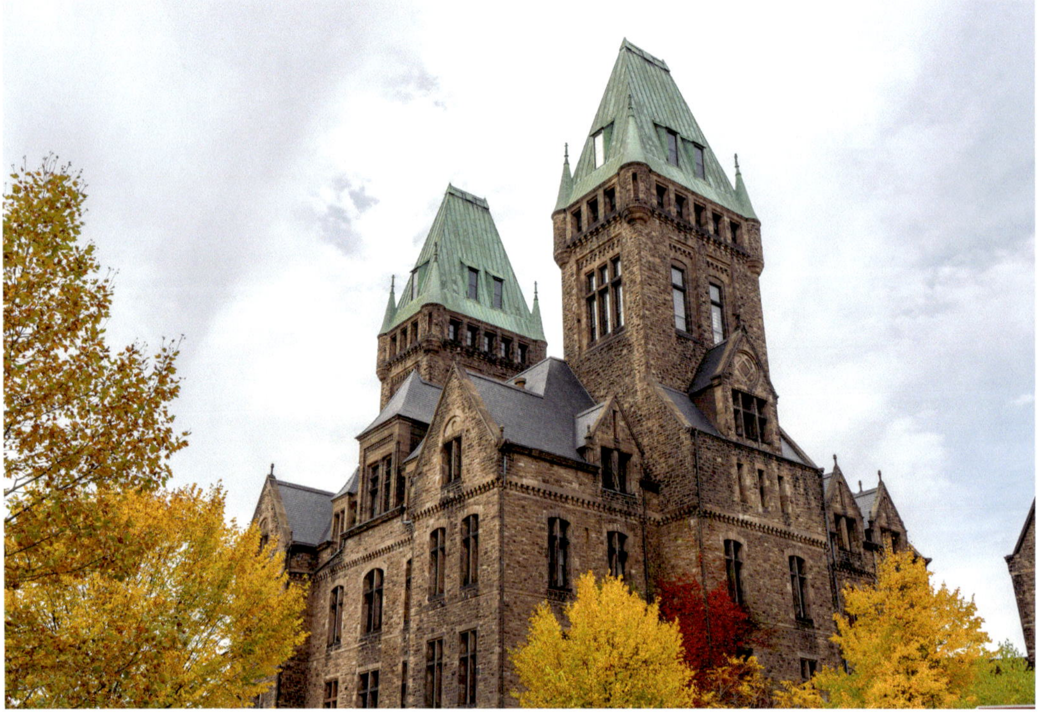

Many Kirkbride asylums were designed to look like castles.

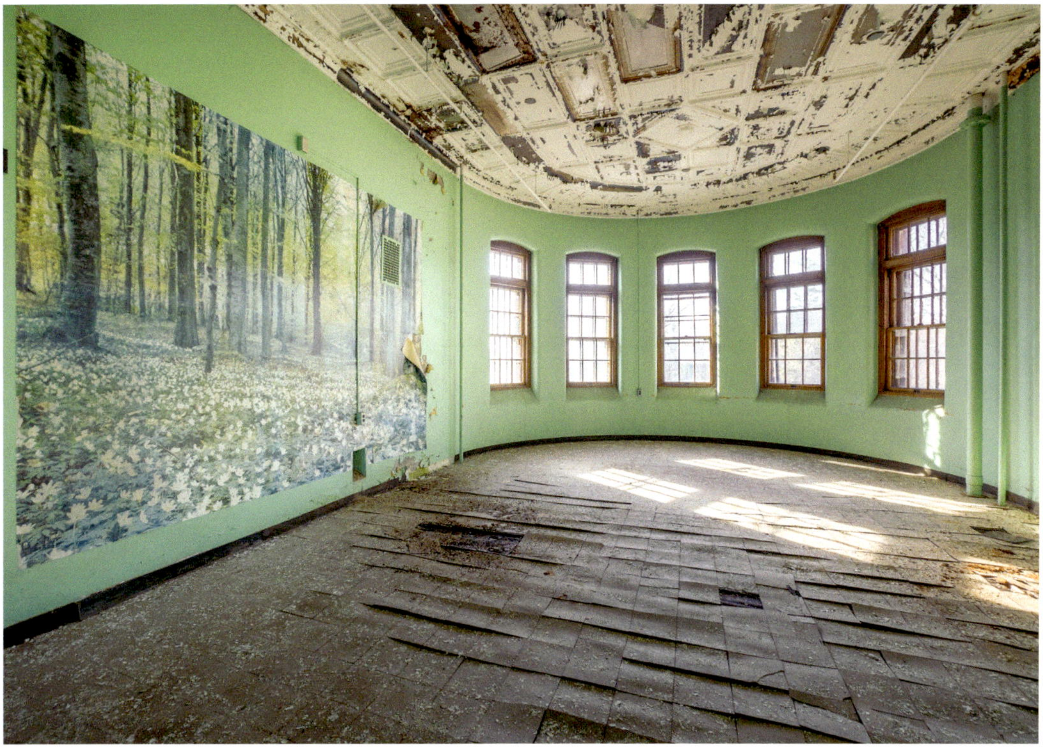

Patient dayrooms often showcase calming color tones such as green, light blue, or pink, along with nature murals.

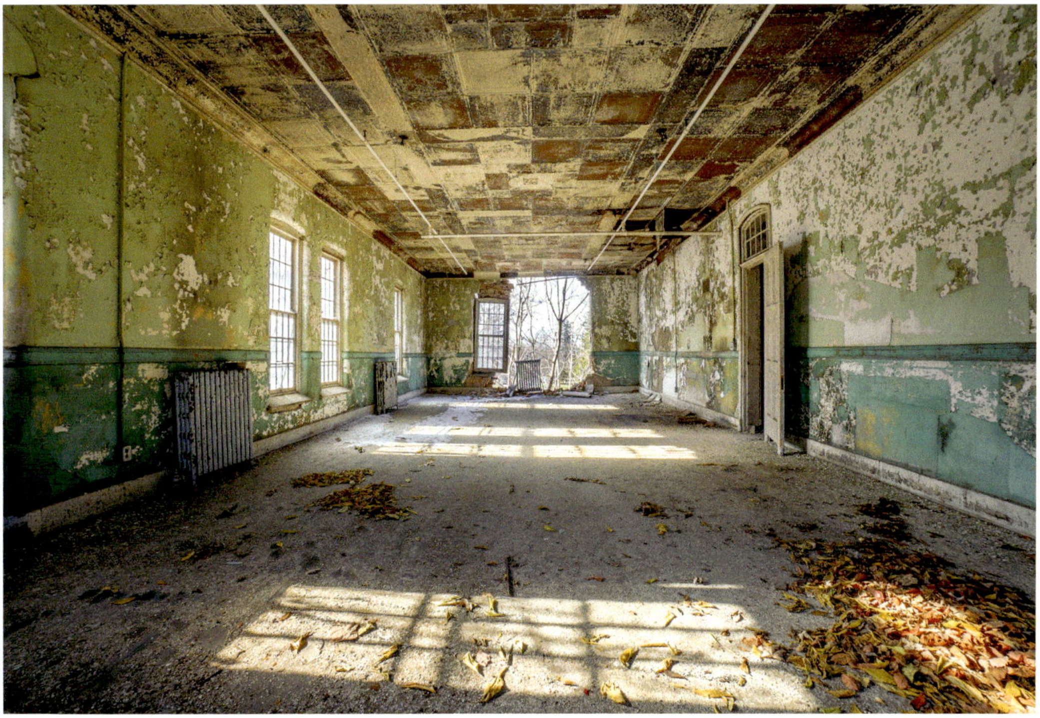

This spacious day room for male patients has ample natural sunlight for reading, watching TV, and socializing.

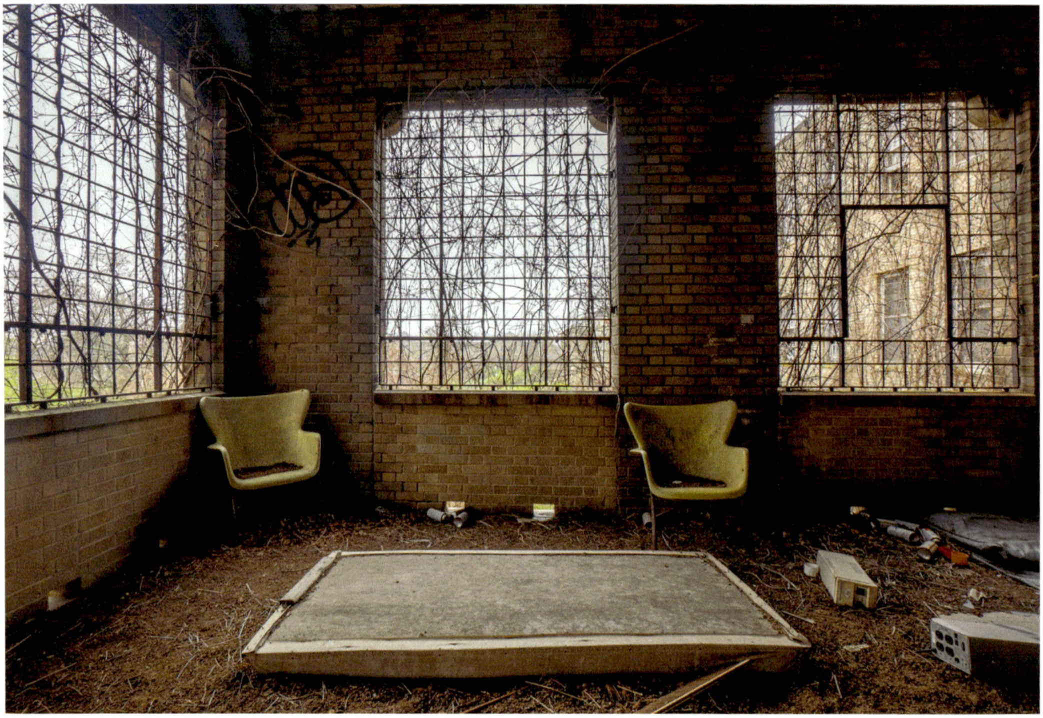

Patient's porches were a great way to get fresh air and sunlight without leaving the hospital.

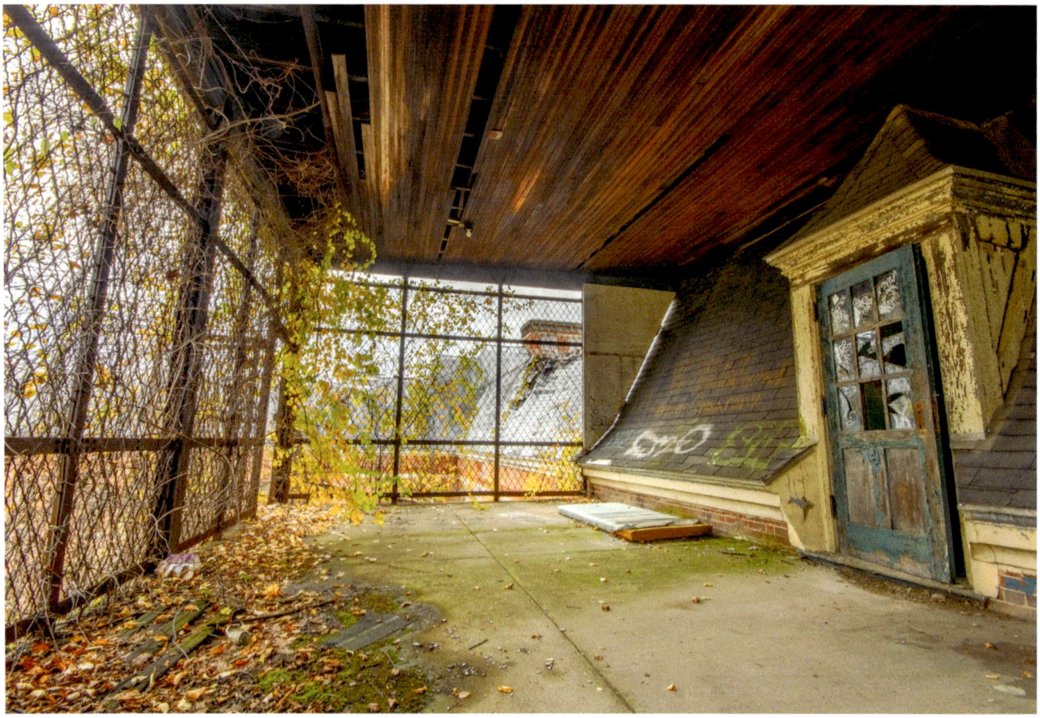

Not only did patients have their porches to relax and enjoy, but so did the staff.

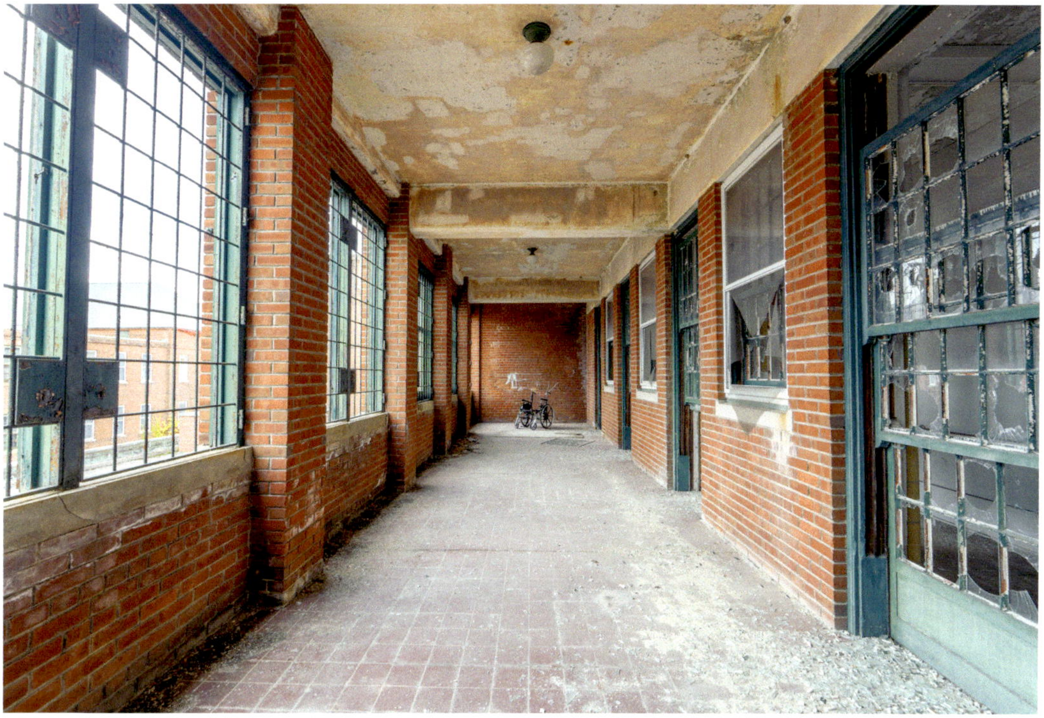

This infirmary ward has constructed slide-up windows that can be used to move bedridden patients to access fresh air and sun.

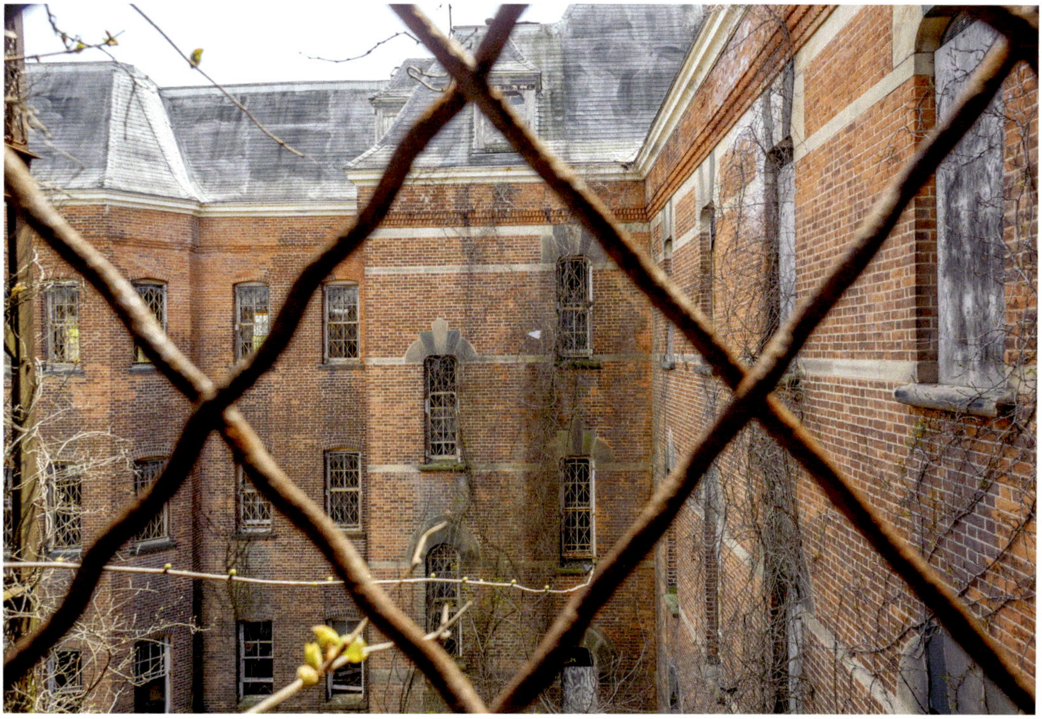

A patient's view from a porch looking at the north wing of a Kirkbride plan hospital.

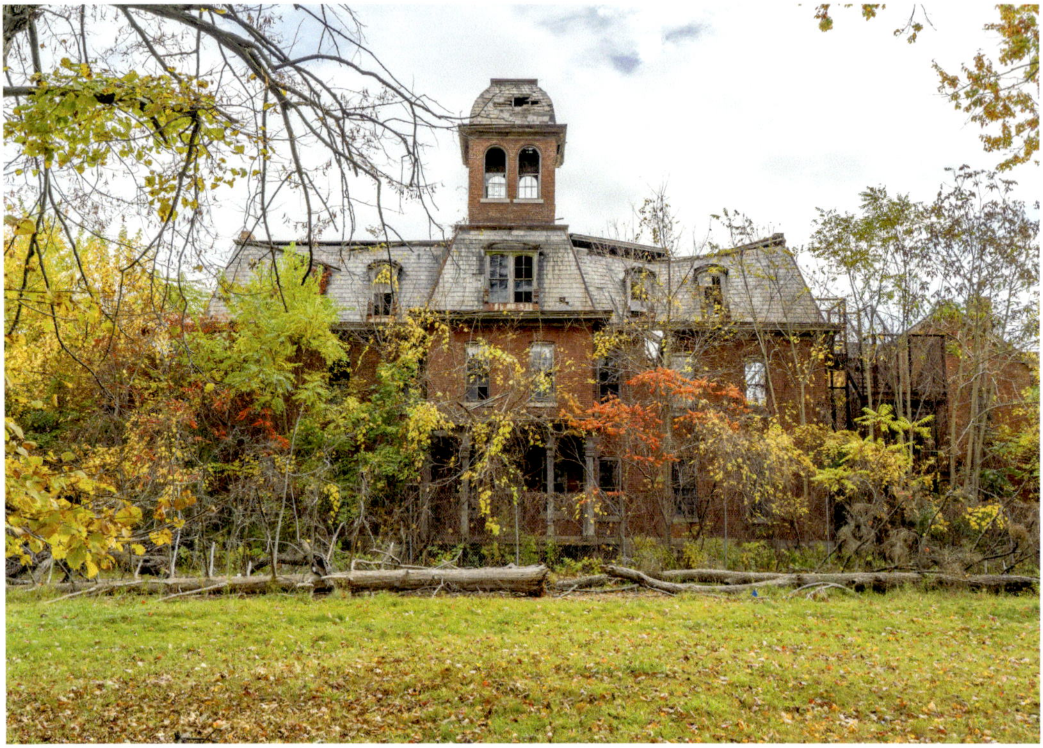

In 1872, the first detached building opened for chronically insane male patients.

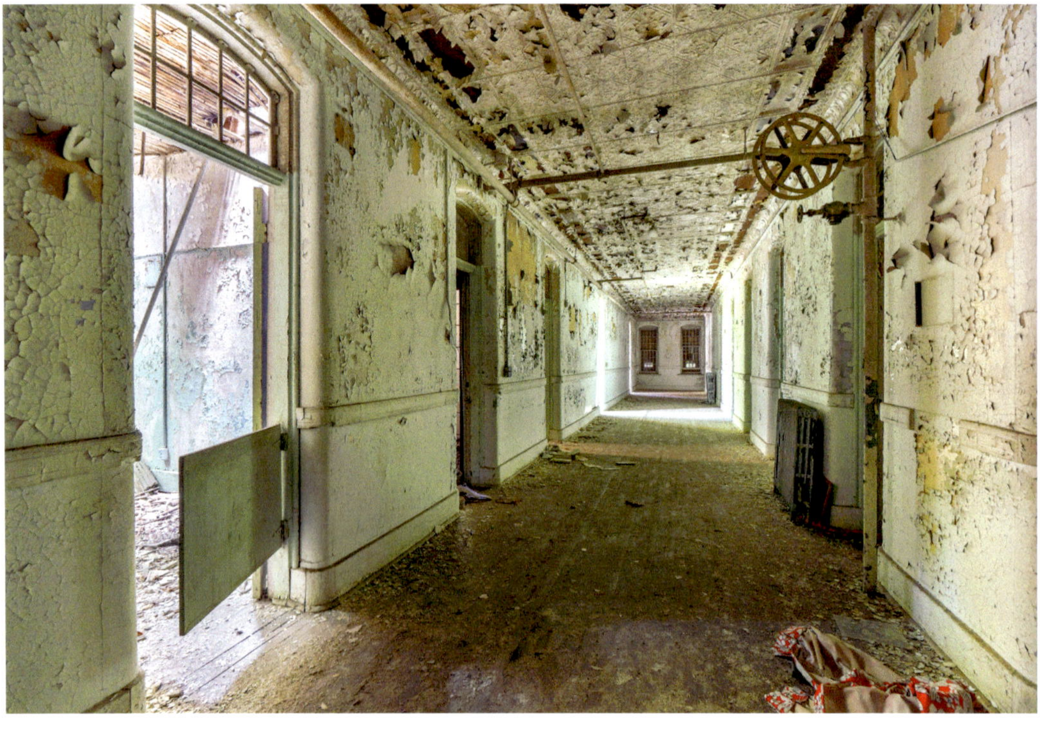

Hallway of the male ward in the second detached building for the chronically insane.

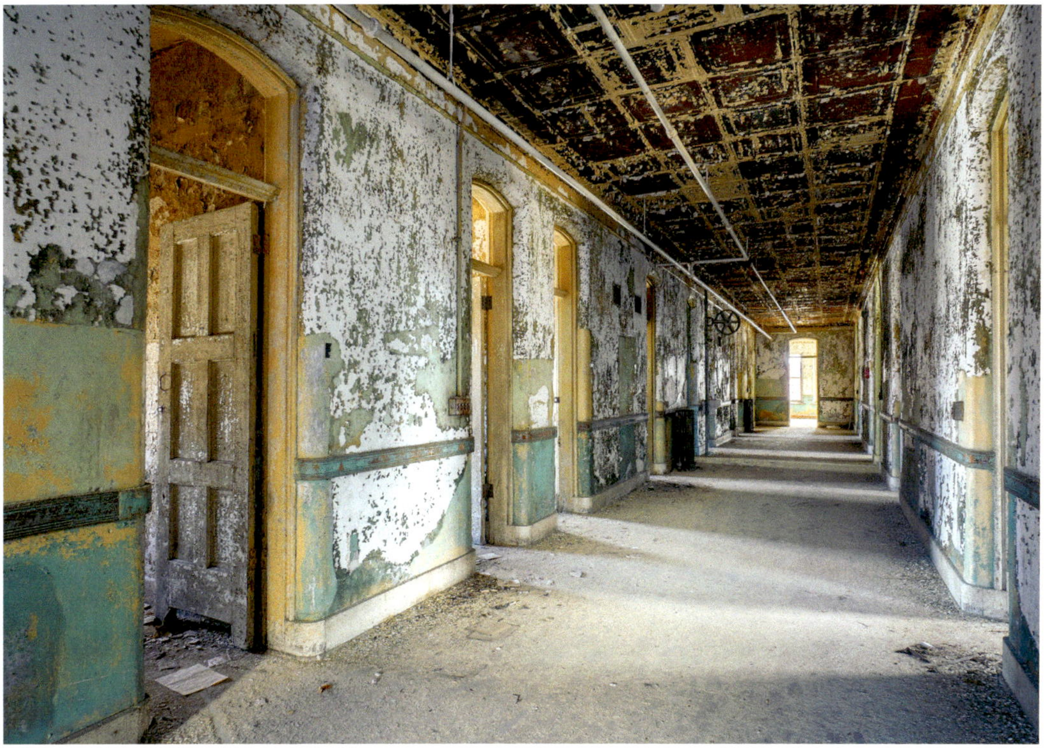

Hallway of the male ward in the first detached building for the chronically insane.

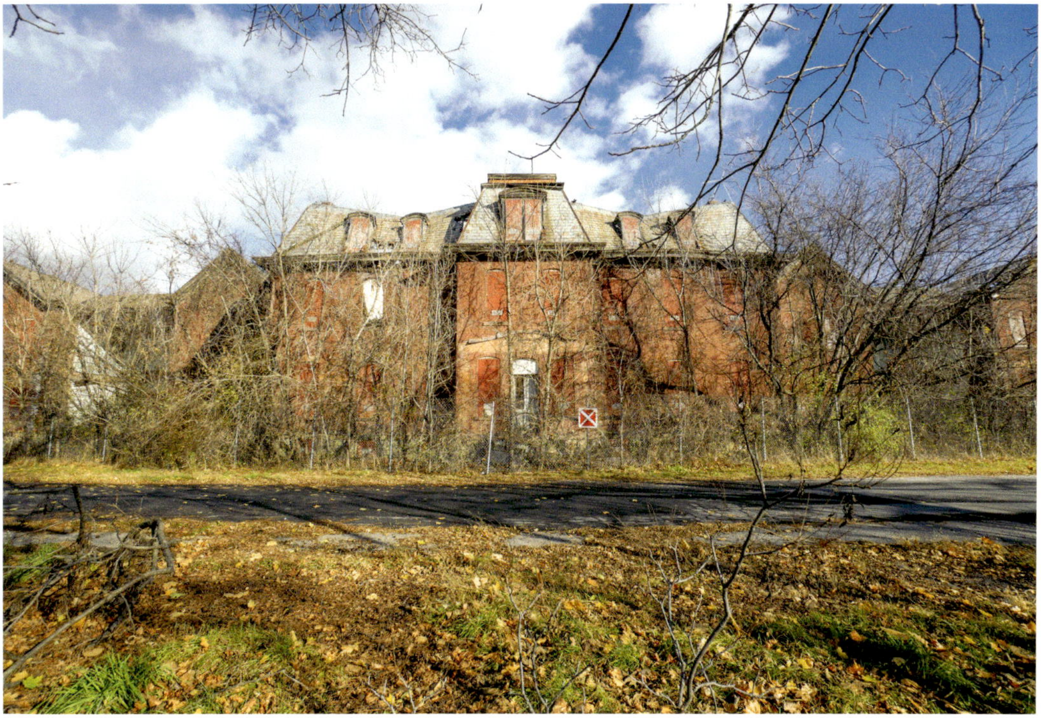

In 1878, the second detached building opened for chronically insane male patients.

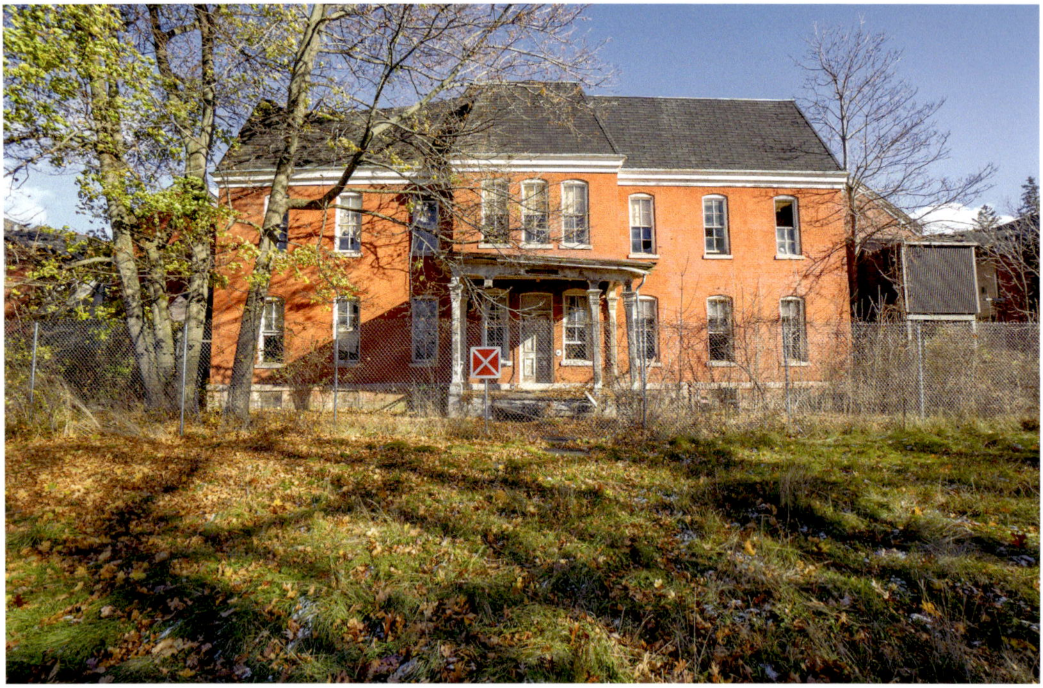

In 1876, the first detached building opened for female patients who were chronically insane.

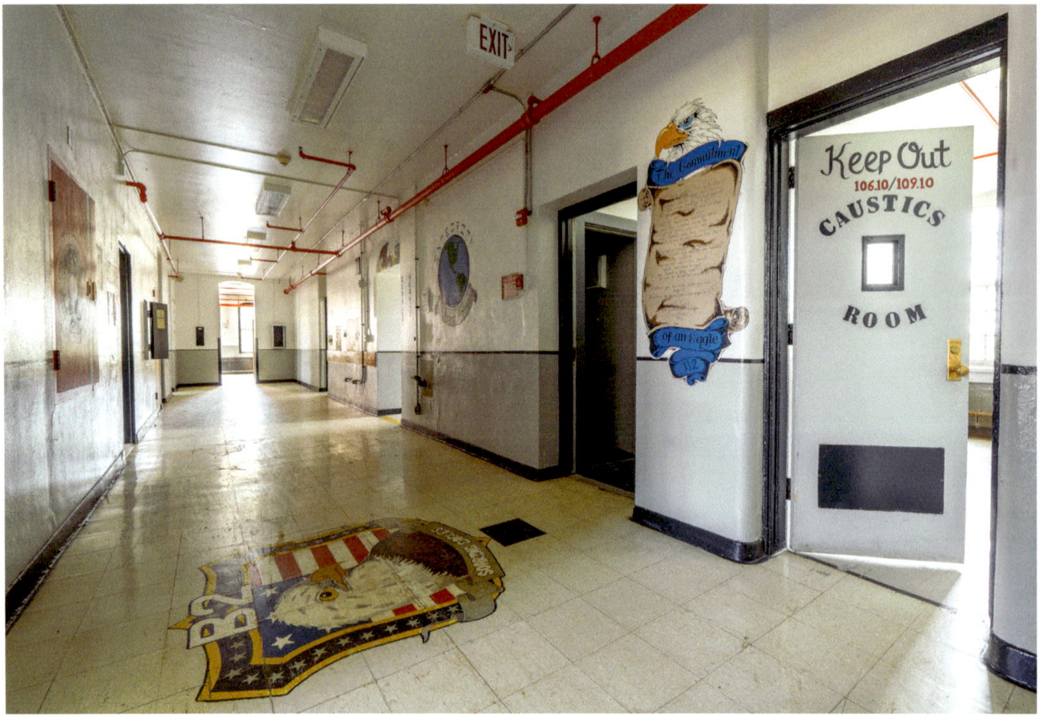

This second detached building's female ward hallway lost all of its original Kirkbride architecture when it was modernized to fit a drug treatment center dormitory.

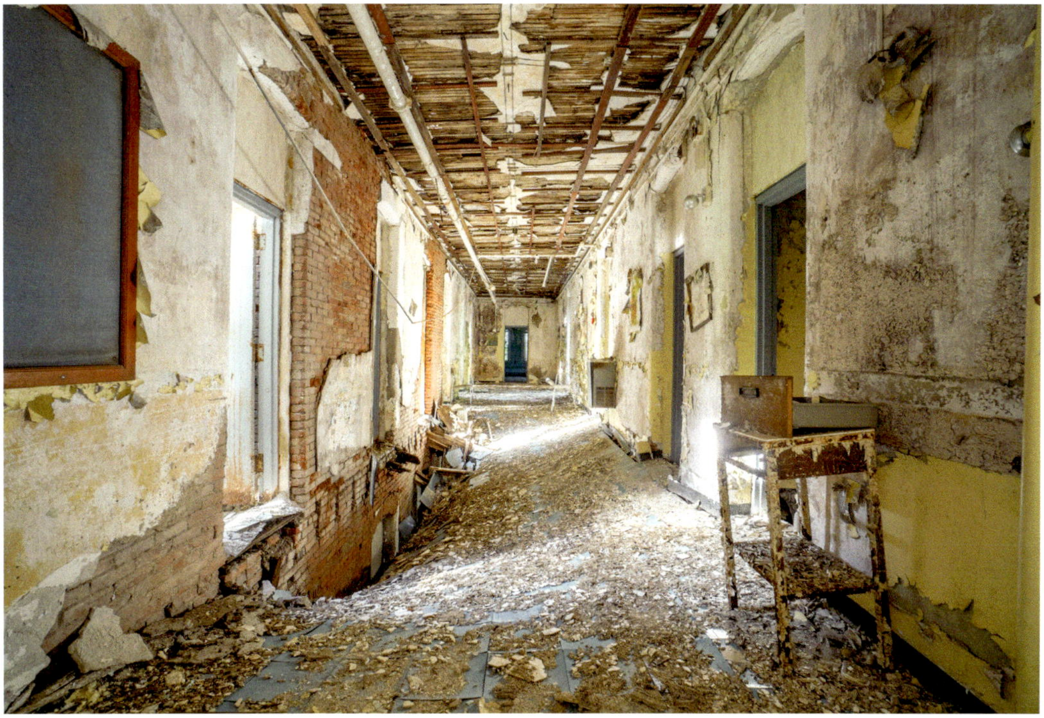

This first detached female building's hallway underwent extensive renovations before it was closed down in the 1990s.

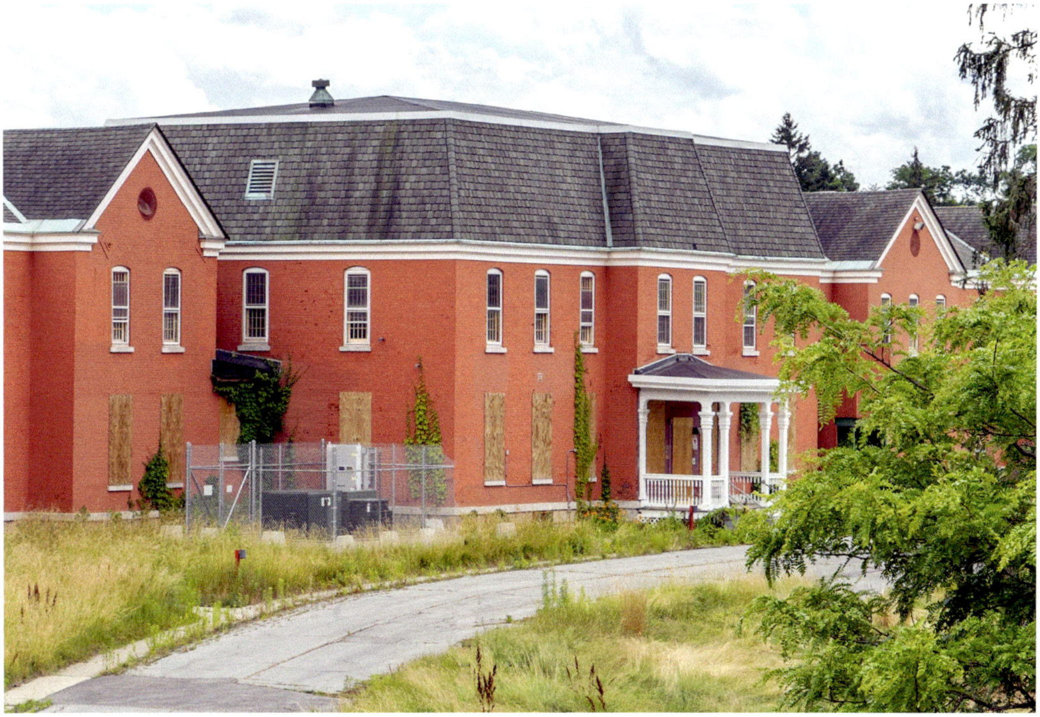

In 1878, the second and last detached building opened for chronically insane female patients.

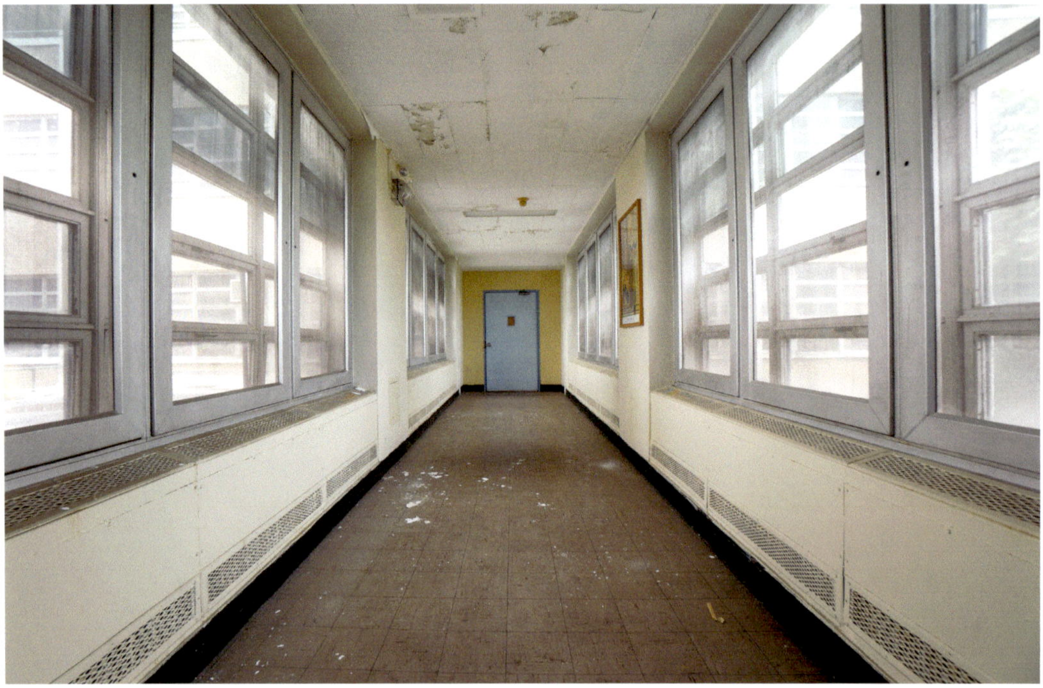

Breezeways were a great way to connect one building to another without ever leaving the hospital.

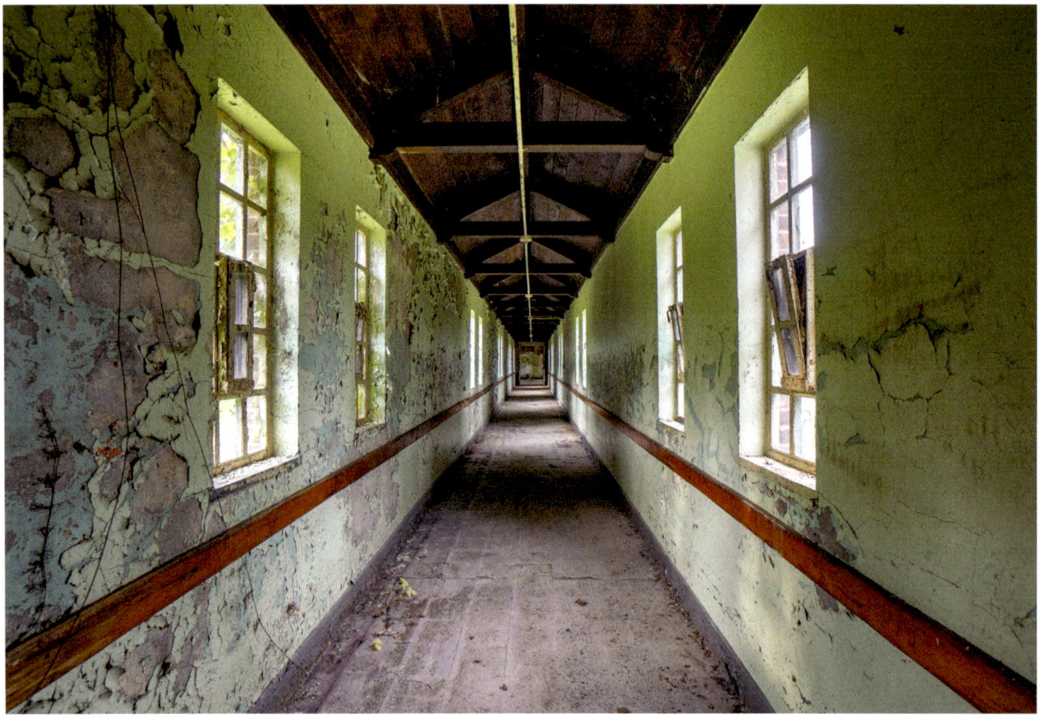

This 1920s breezeway connects the patient's ward to the main dining hall.

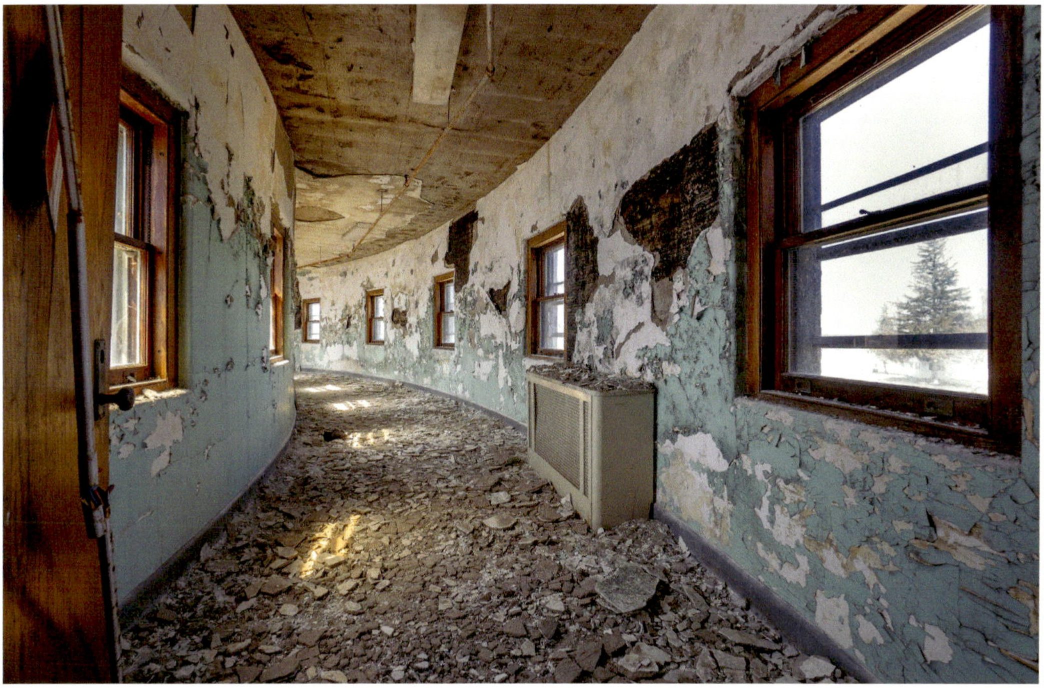

This hospital added a curved breezeway to connect the upper patient wards, improving accessibility between each cottage.

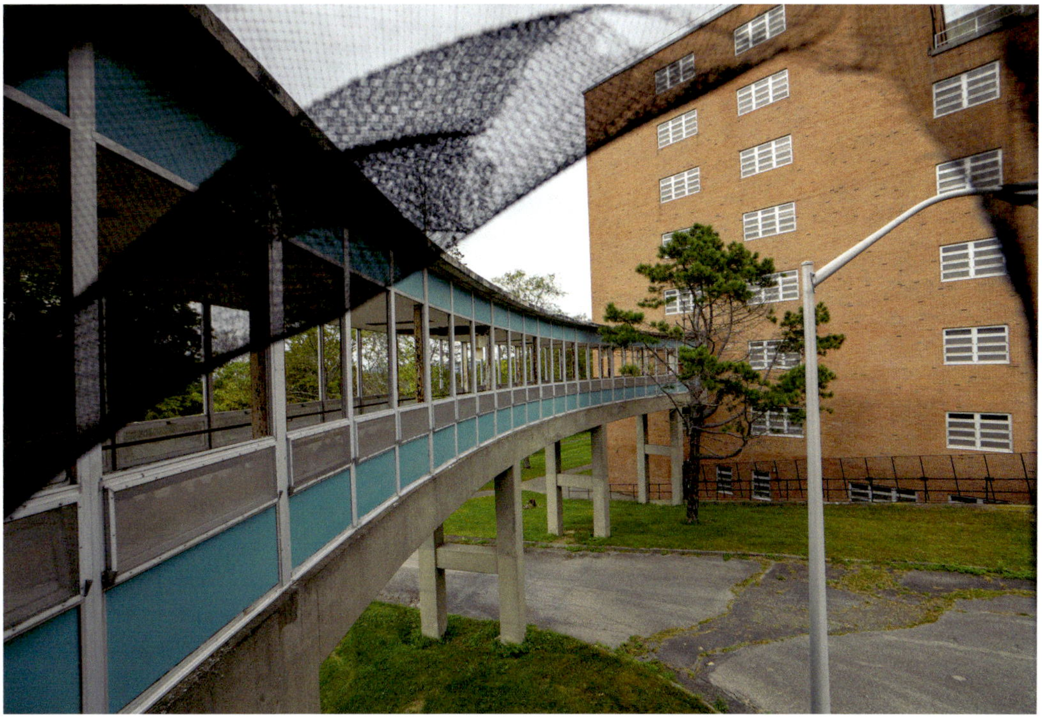

This arched medical breezeway connected an older hospital building to a newer one, aiding both patients and doctors.

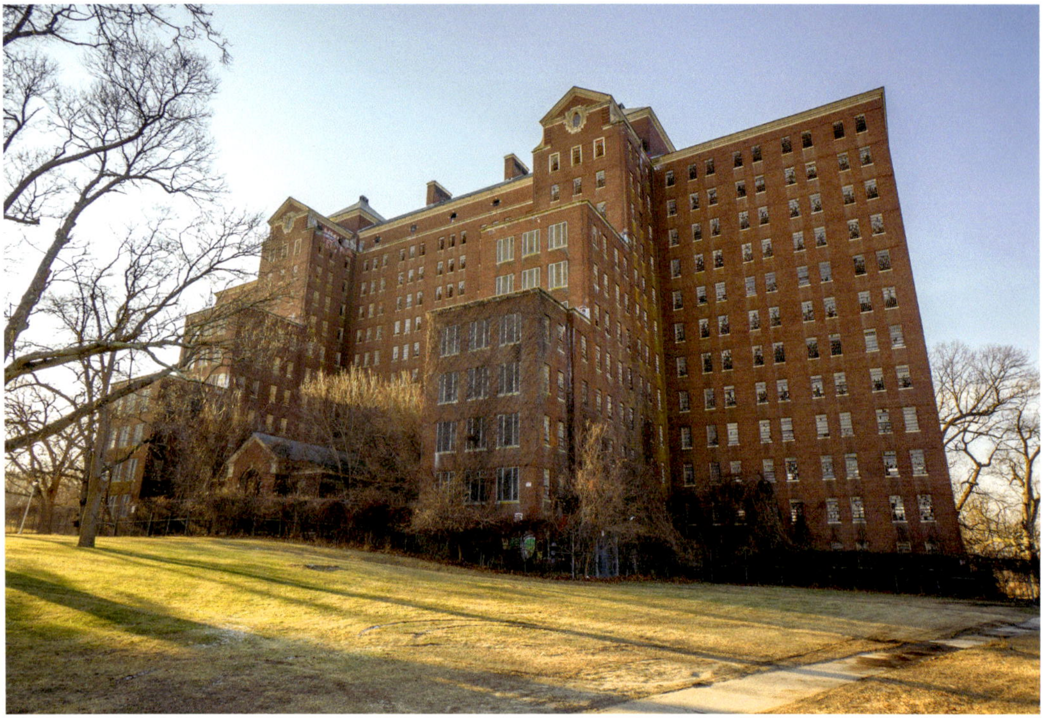

Around the 1930s, New York started to build massive high-rises to house its growing patient population.

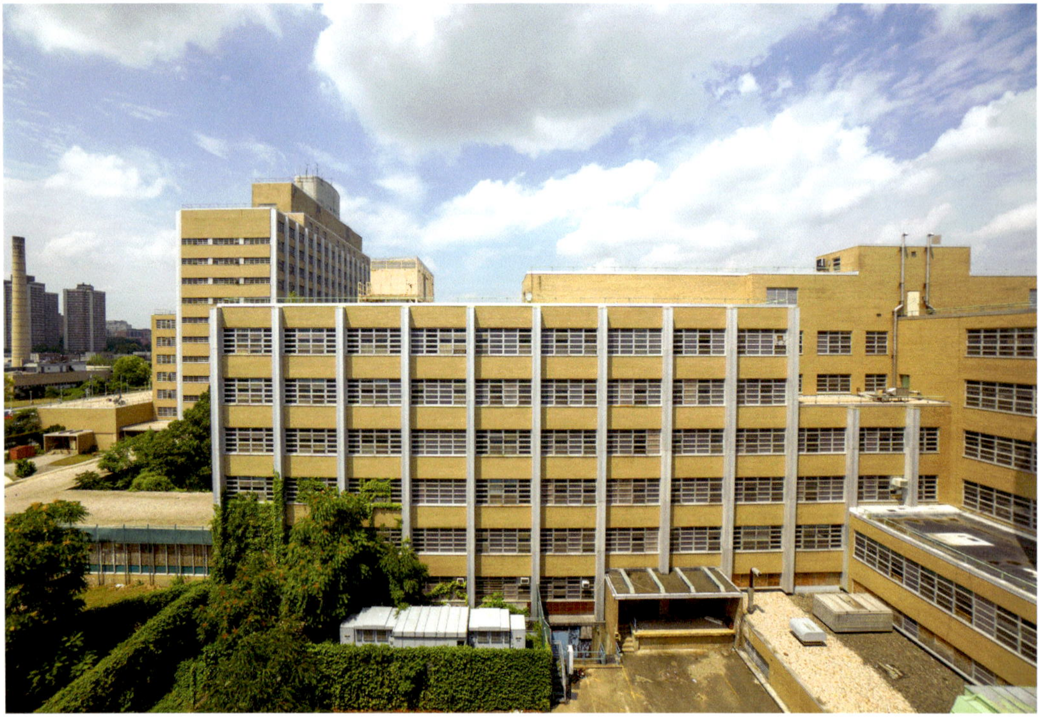

This 1960s high-rise institution could house up to 680 patients in one single building.

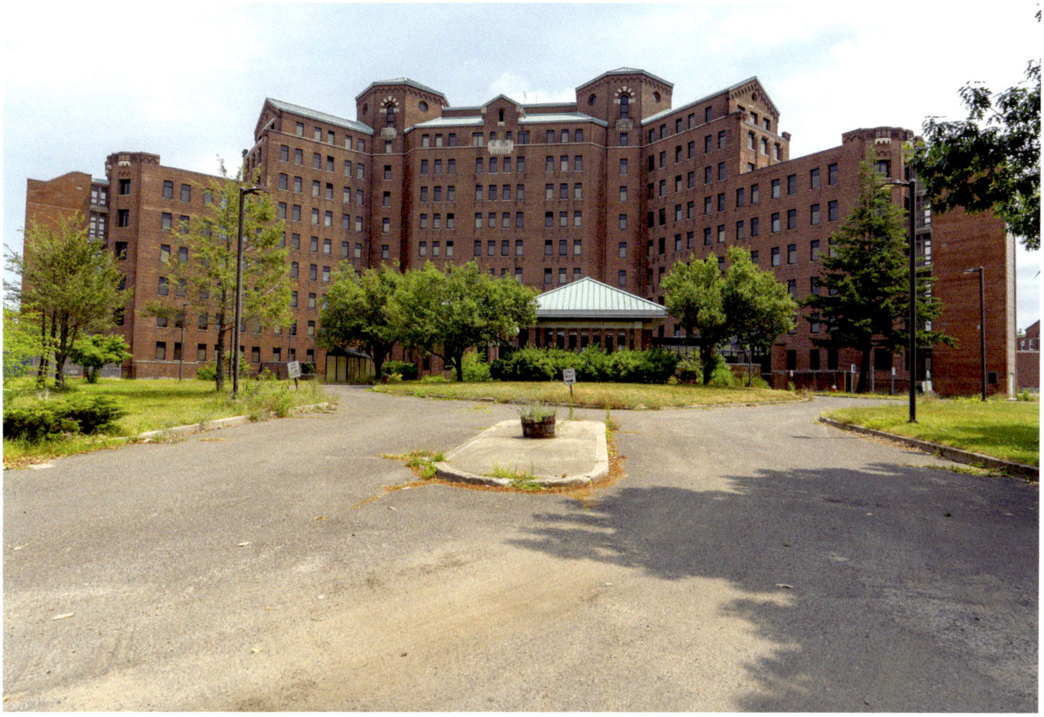

Most of these new massive structures contained a medical department, dental clinic, therapy department, patient wards, and a morgue, replacing the older buildings that were outdated.

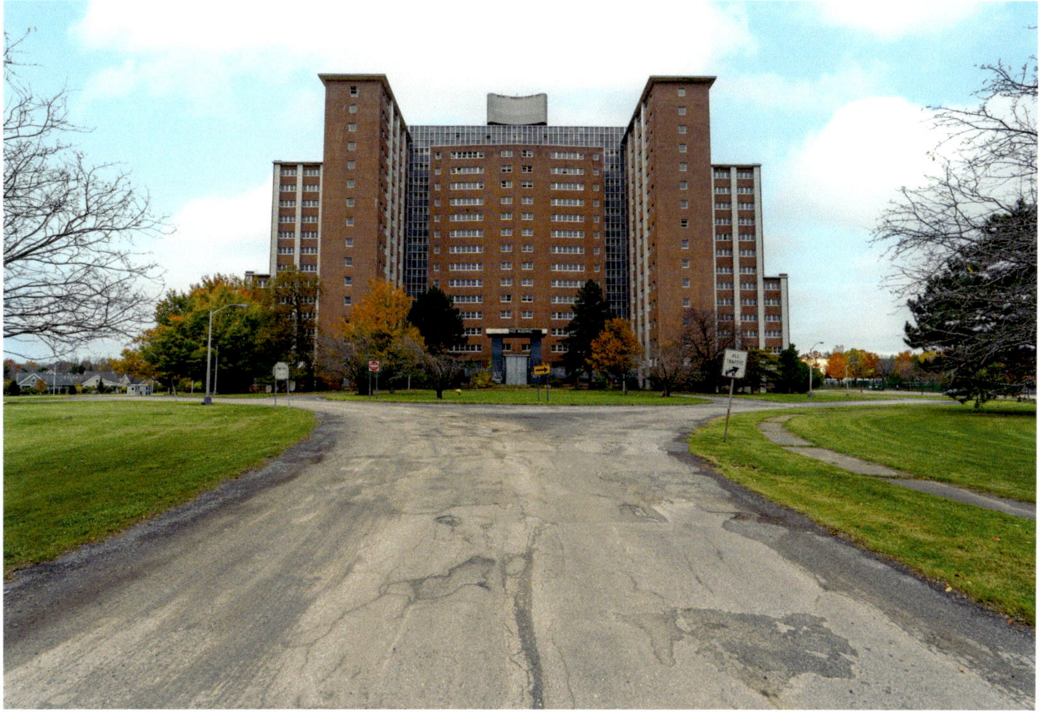

Deinstitutionalization in the 1950s prevented this high rise from being fully utilized.

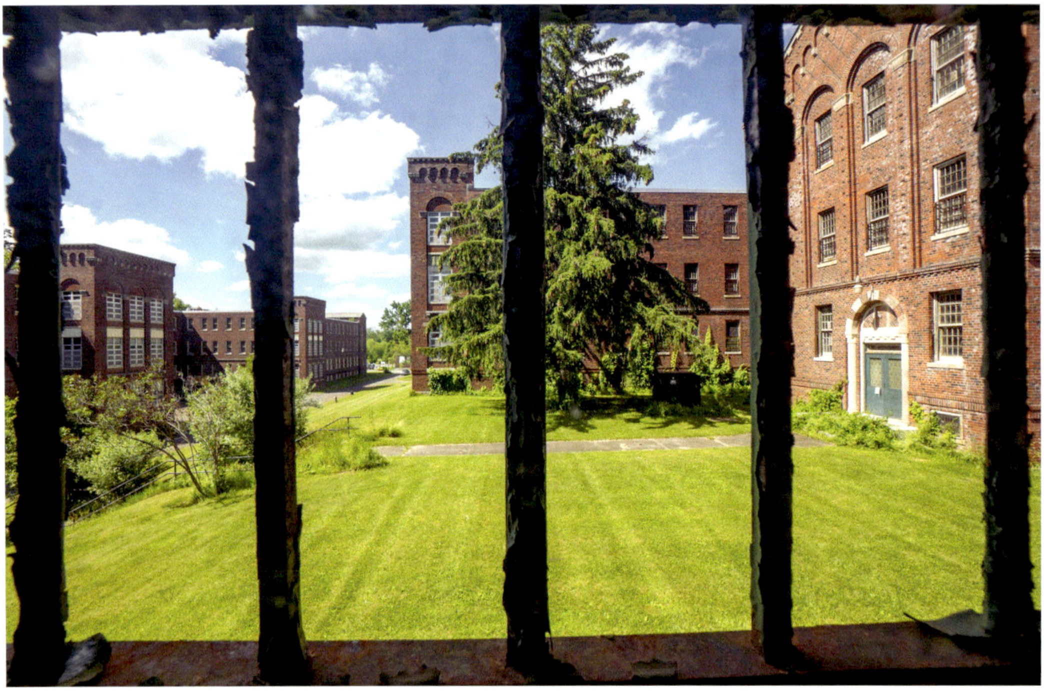

Some institutions were designed to look and feel more like beautiful college campuses than large institutionalized housing for the insane.

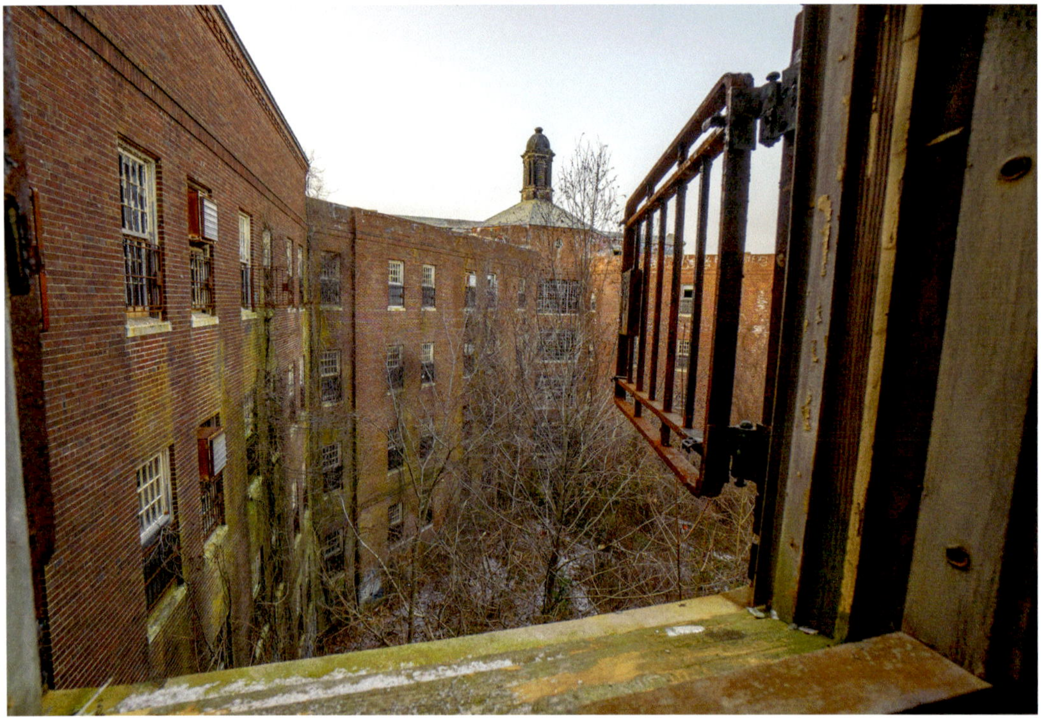

Over time, asylums replaced their decorative window grates with bar-like designs.

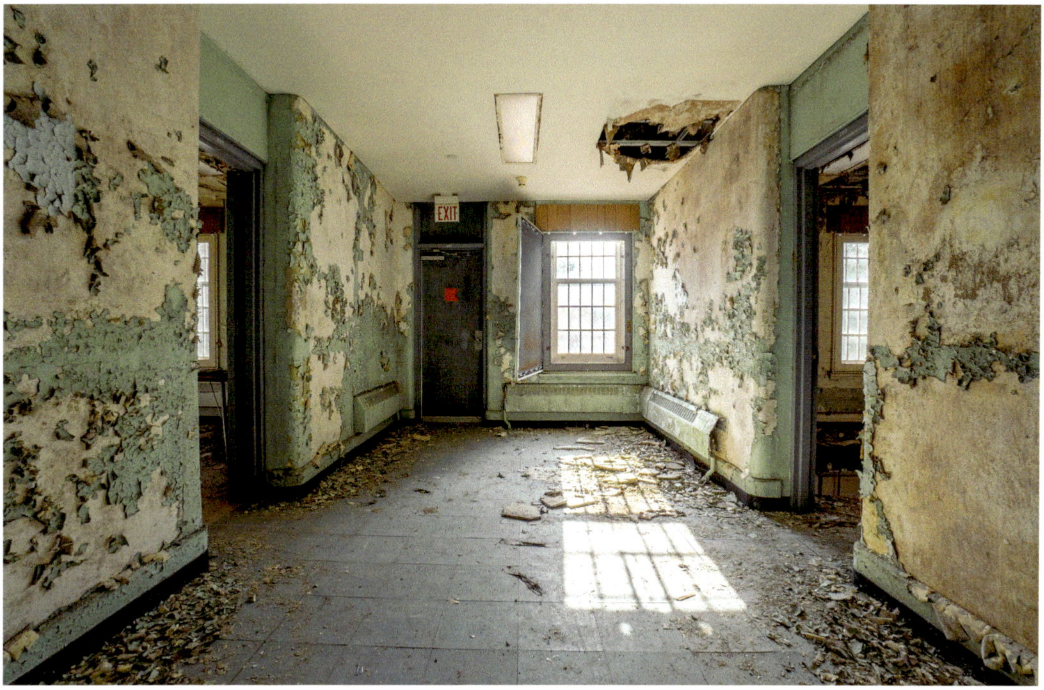

Modern renovations have obscured the original beauty of this Kirkbride asylum hallway, hiding its tin ceilings, large glass windows, wooden doors, and floors.

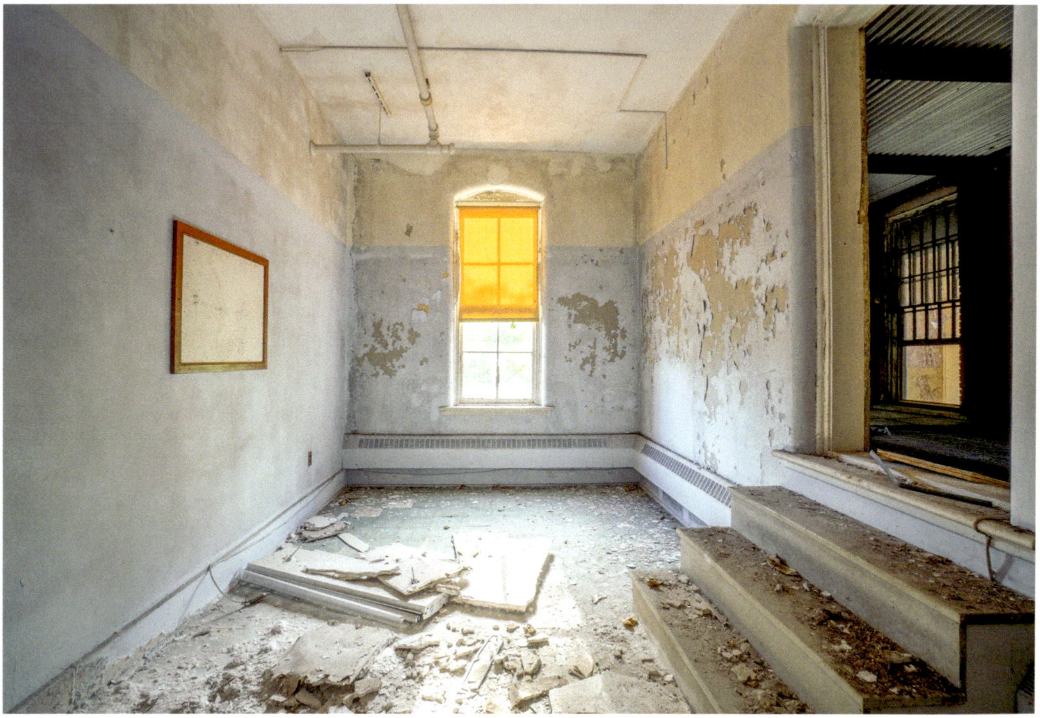

Other renovations included the implementation of new safety standards, such as the addition of fire escapes where necessary.

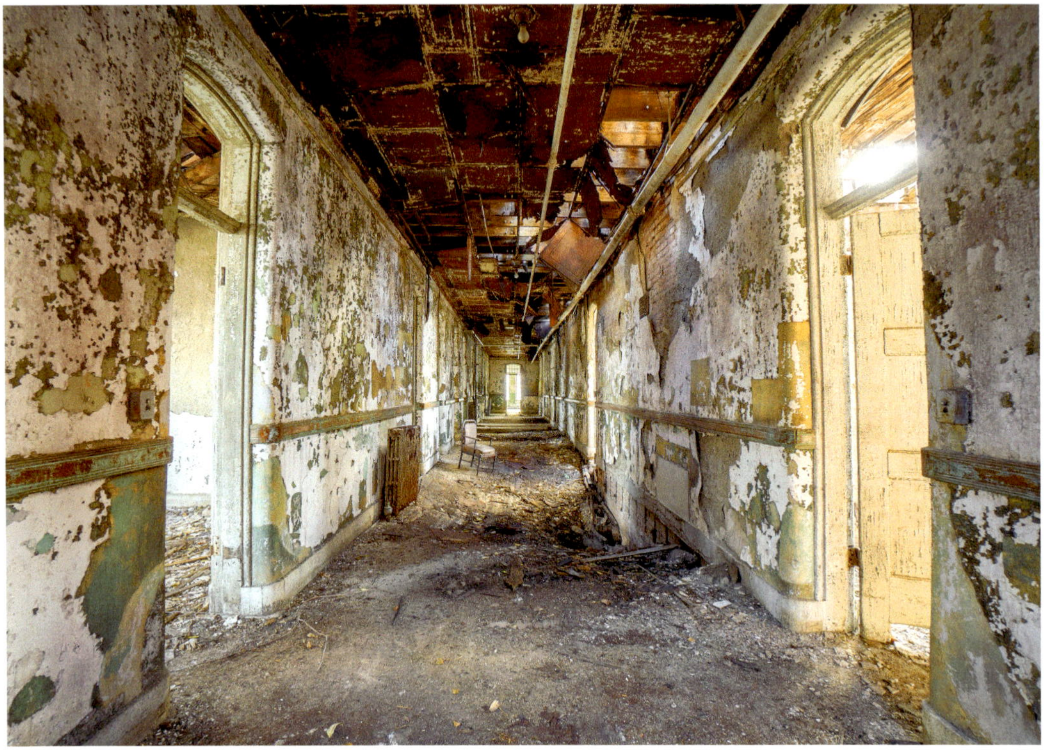

This detached male building held eight 50-bed dormitories for patients, with a kitchen, offices and dayrooms.

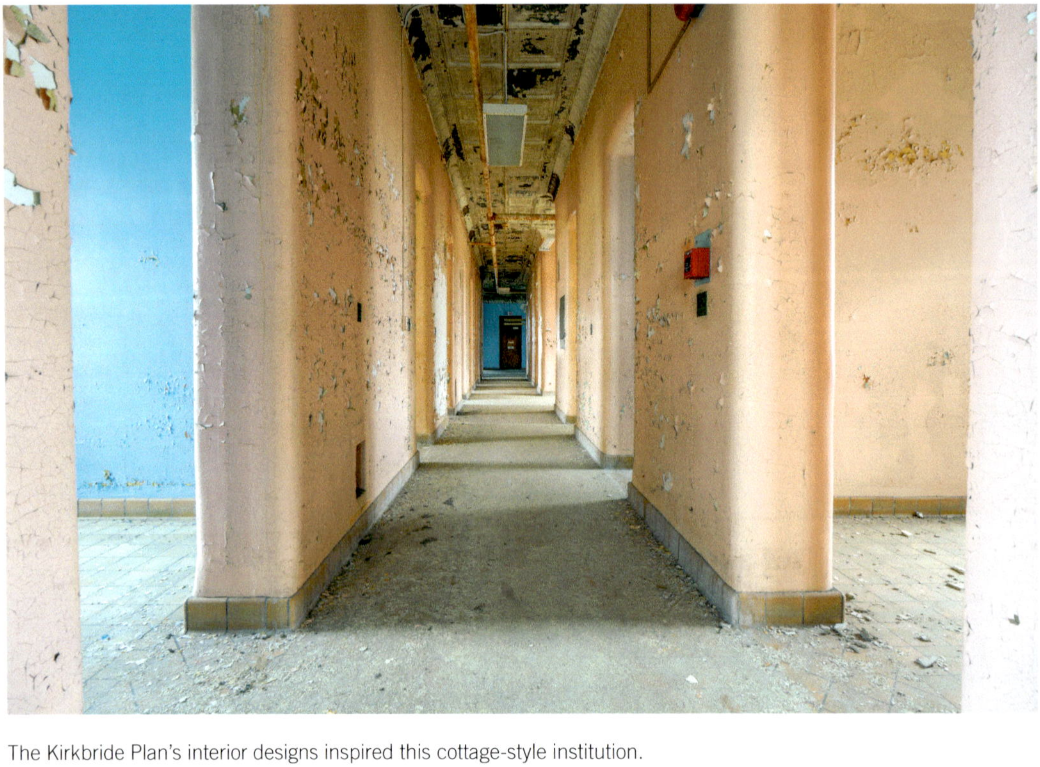

The Kirkbride Plan's interior designs inspired this cottage-style institution.

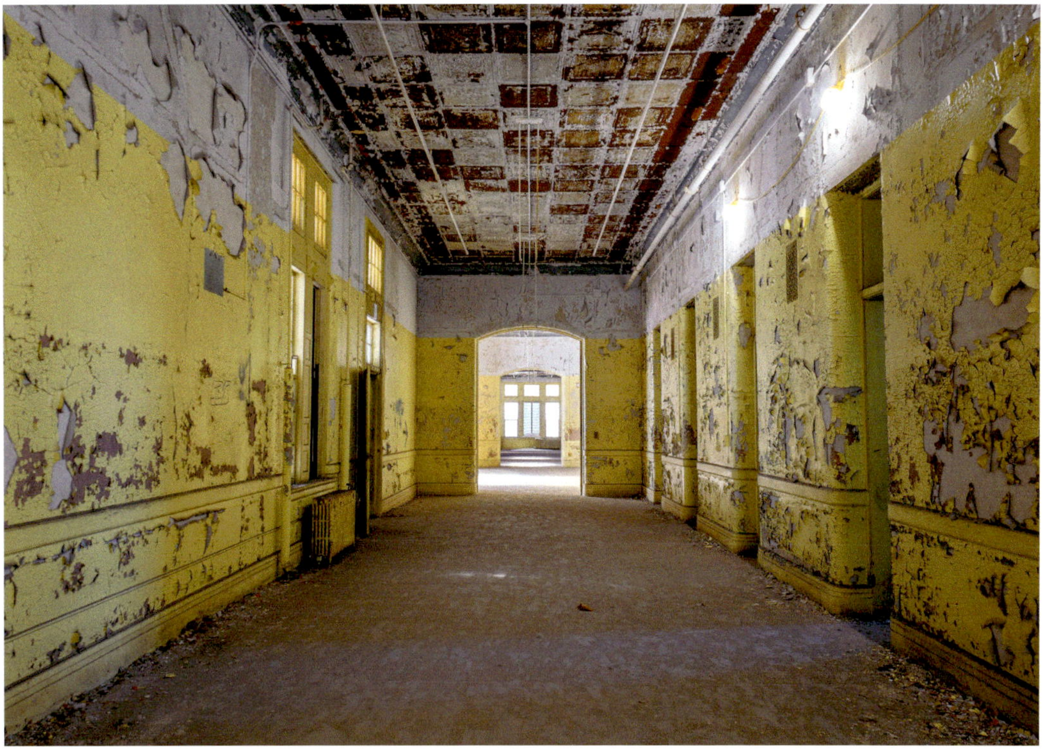

This Kirkbride institution was designed by famous architect Henry Hobson Richardson.

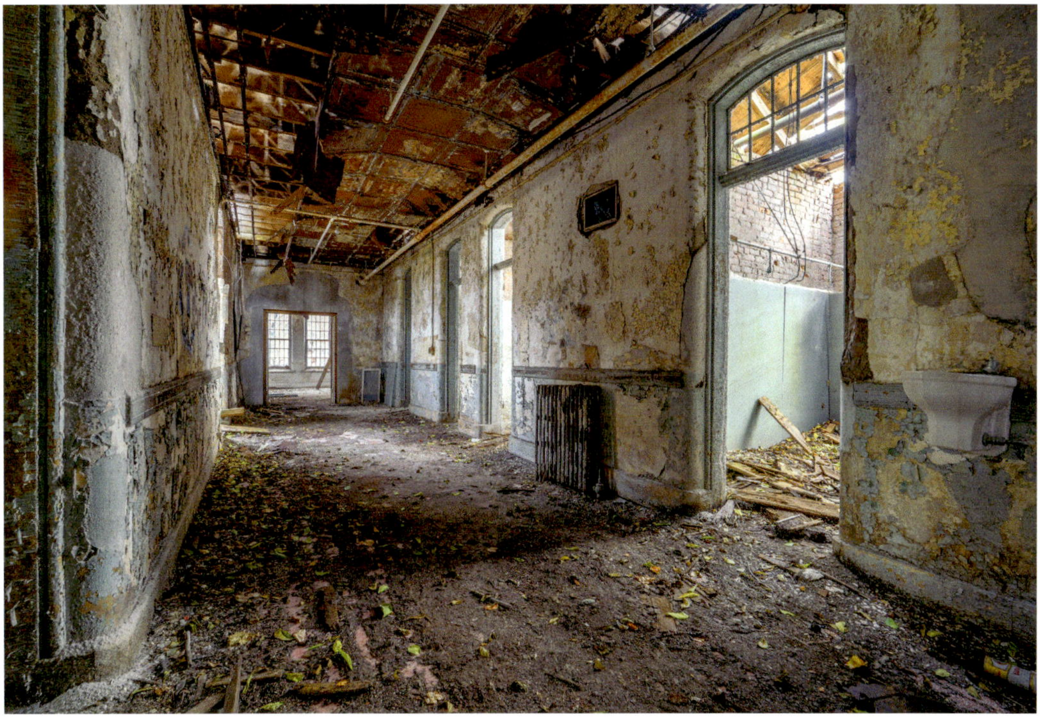

Transom windows on top of doors were important for allowing the passage of air and light.

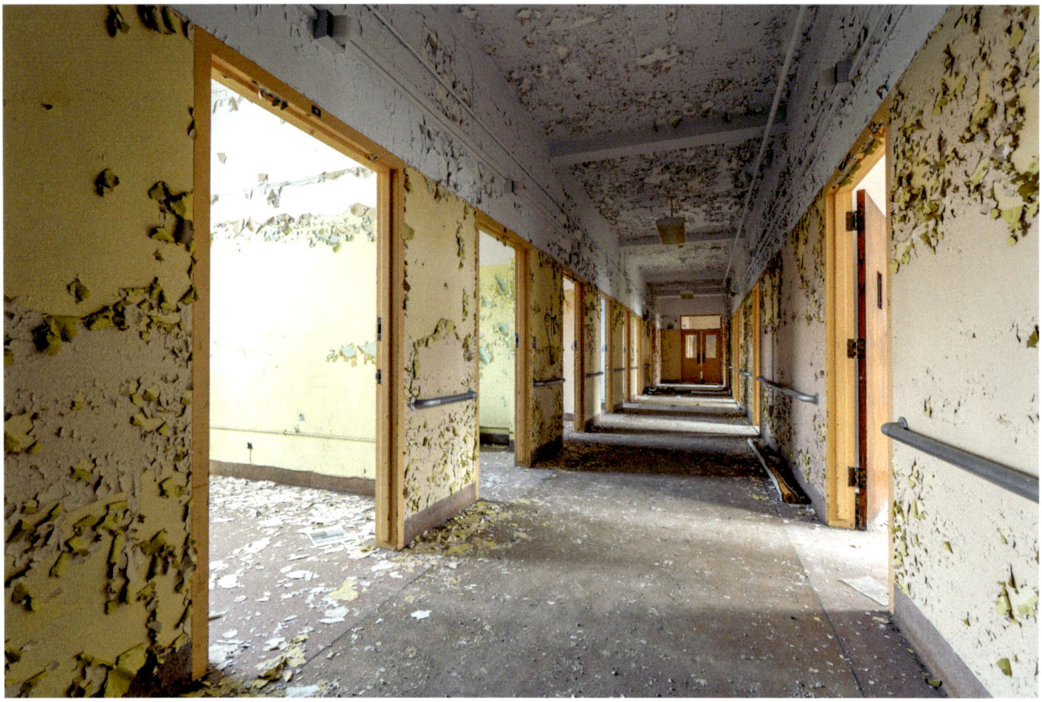

In the 1930s, with new treatments and modern technologies, the once beautiful asylum hallways began to simplify, losing the intricate details they were famous for.

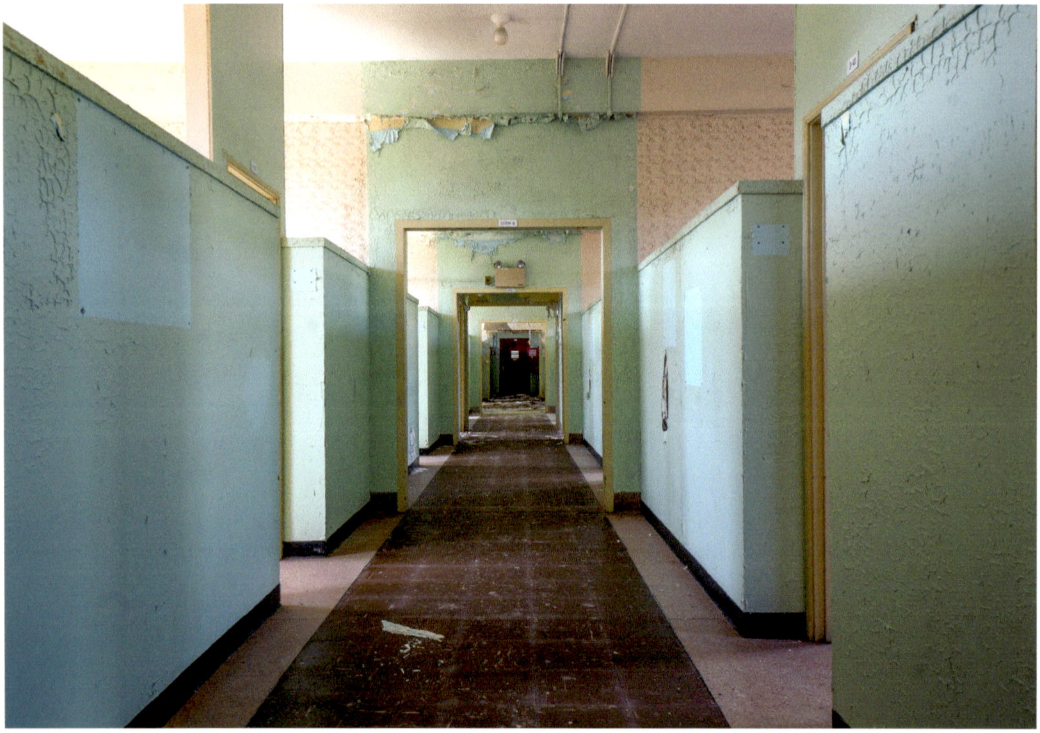

To improve patient privacy, walls, and cubicles were constructed to replace open wards.

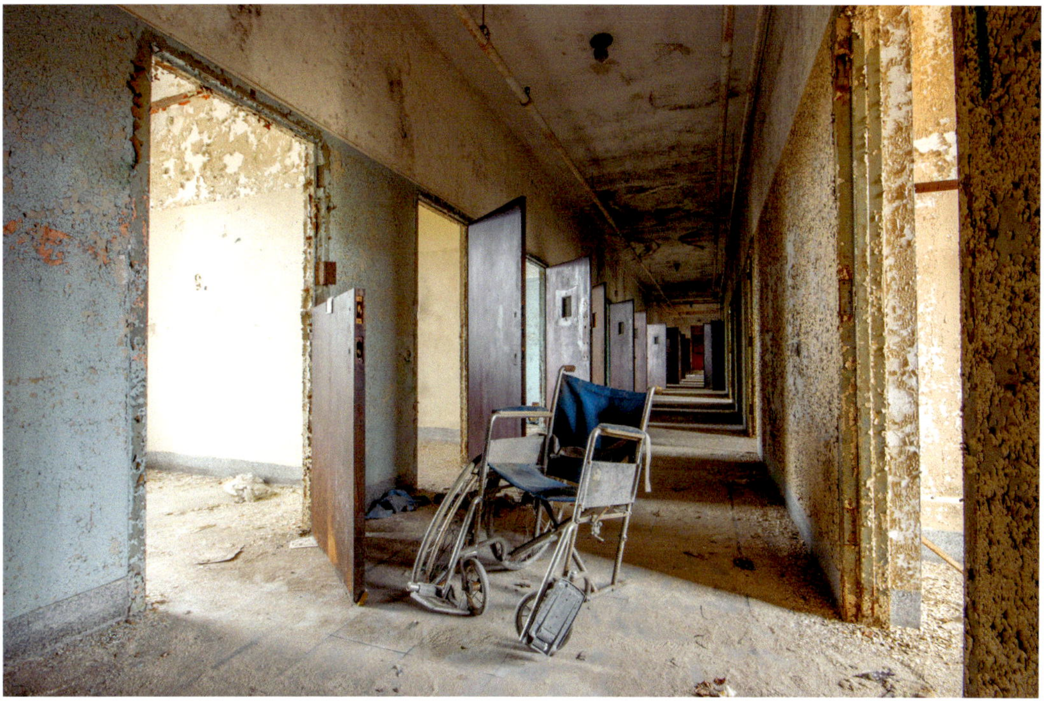

Most post-war asylums in New York were constructed with concrete slab ceilings, asbestos tiles, and heavy wooden doors.

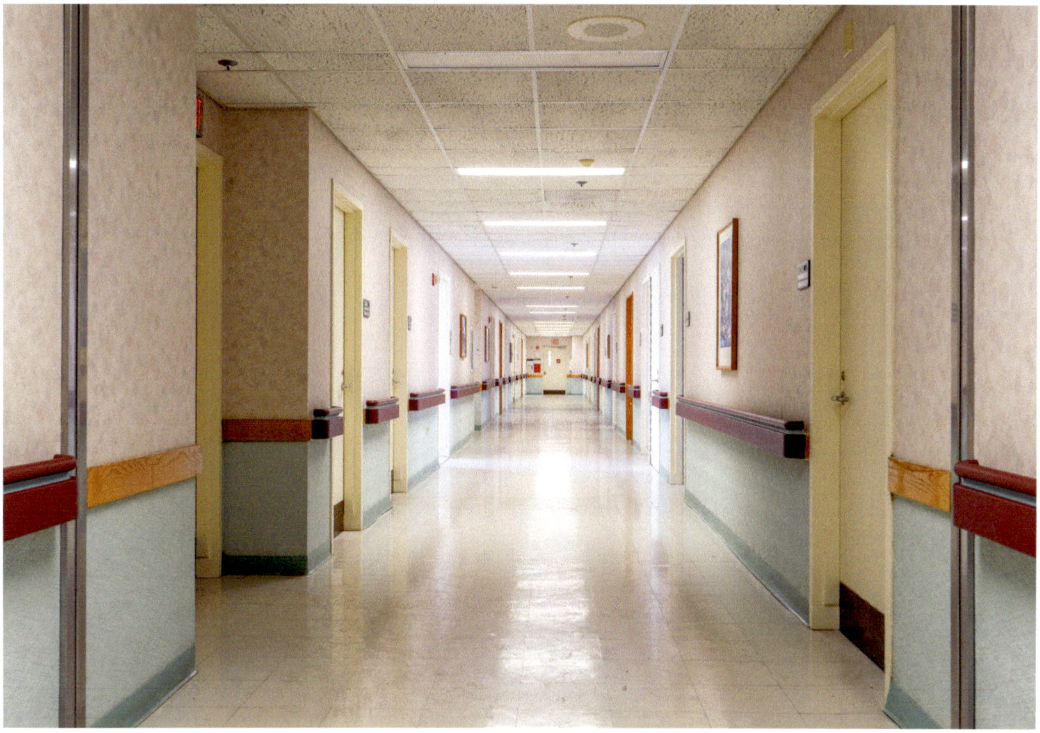

This renovated patient ward hallway illustrates the contrast between modern and early institutions designs.

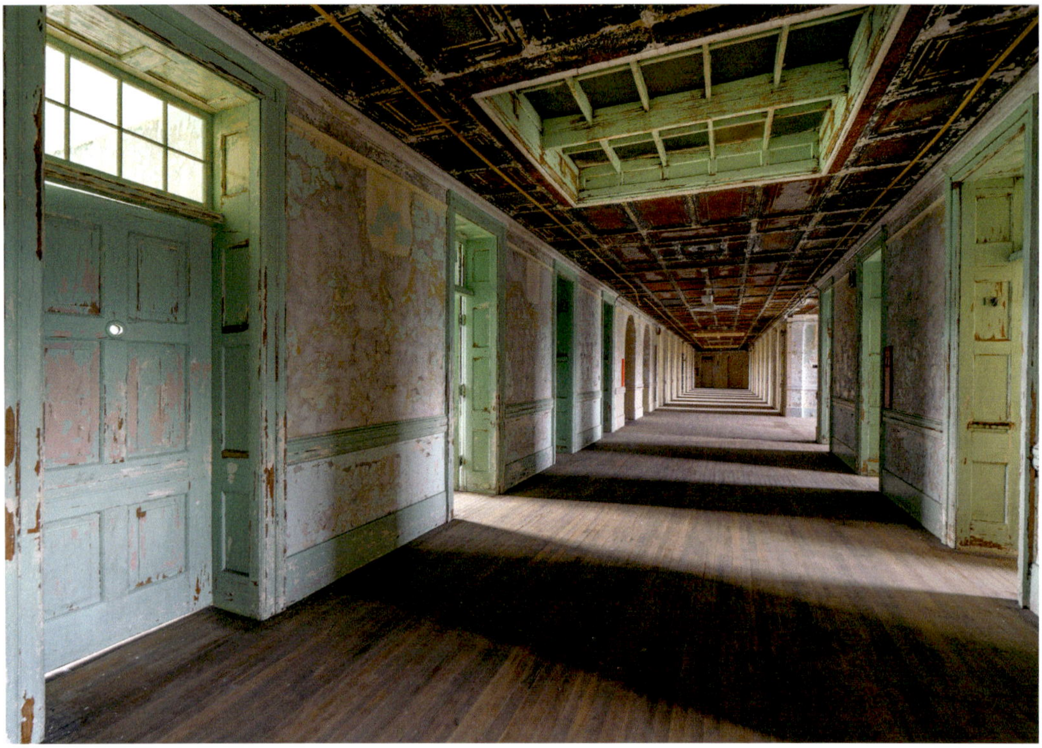

Natural light was crucial, this second-floor patient's ward received sunlight from a rooftop window from a floor above.

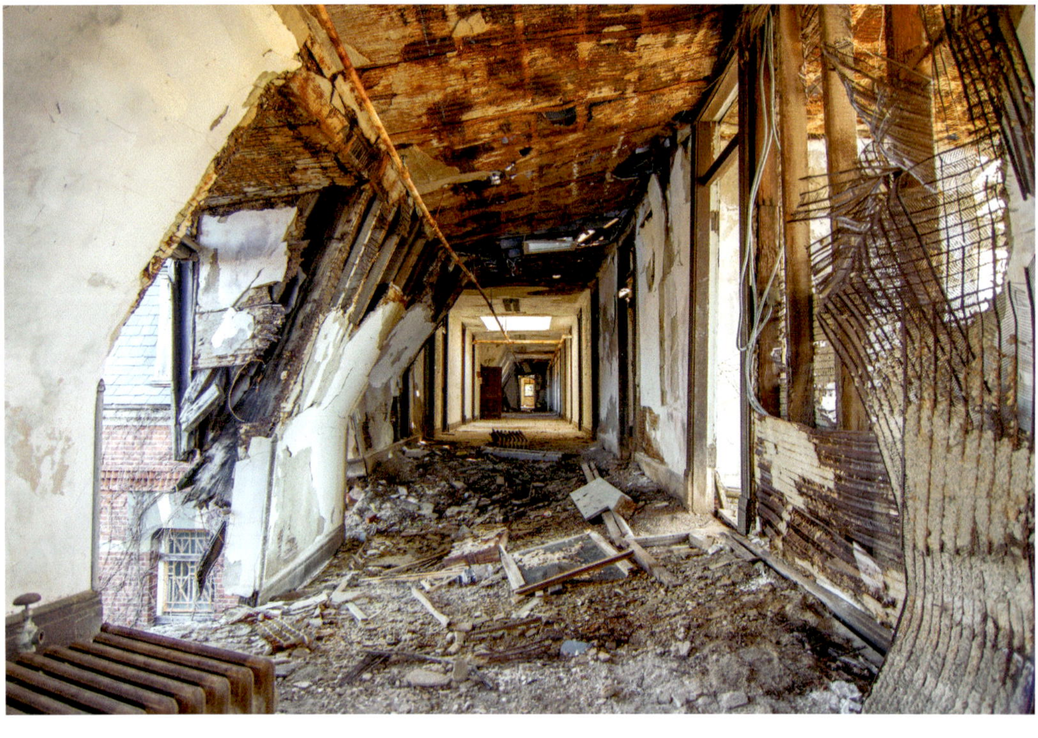

In some early institutions, the staff lived on the upper floors above the patients.

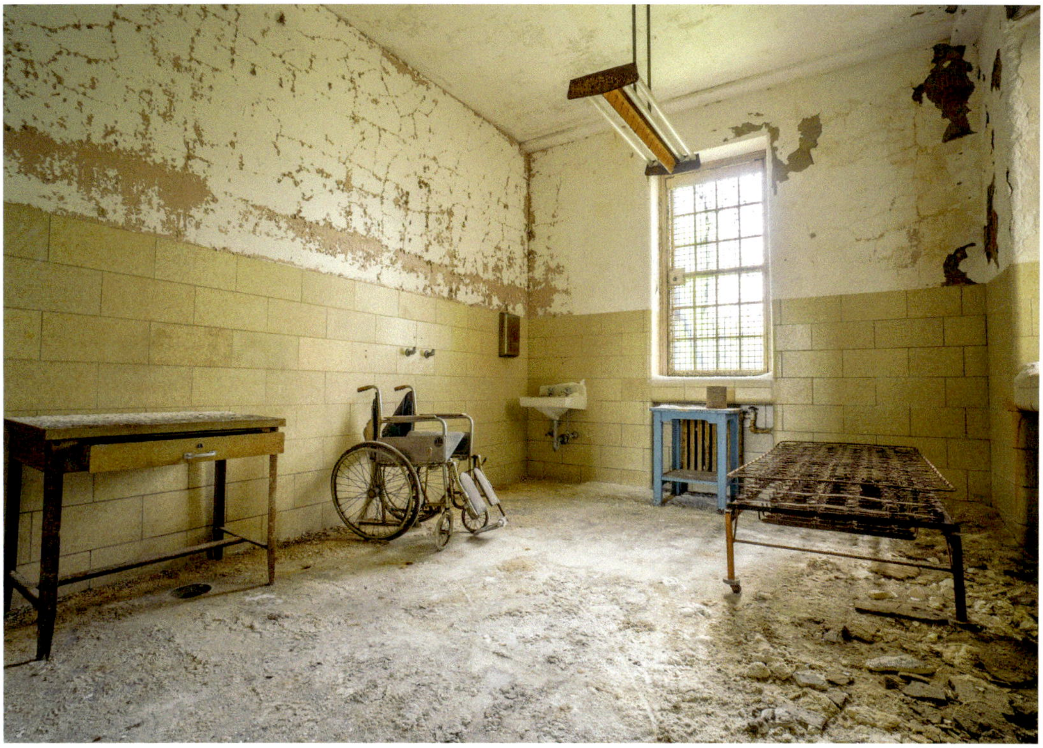

This semi-gloss yellow brick tile became popular in the 1930s to help with noise control and hygiene.

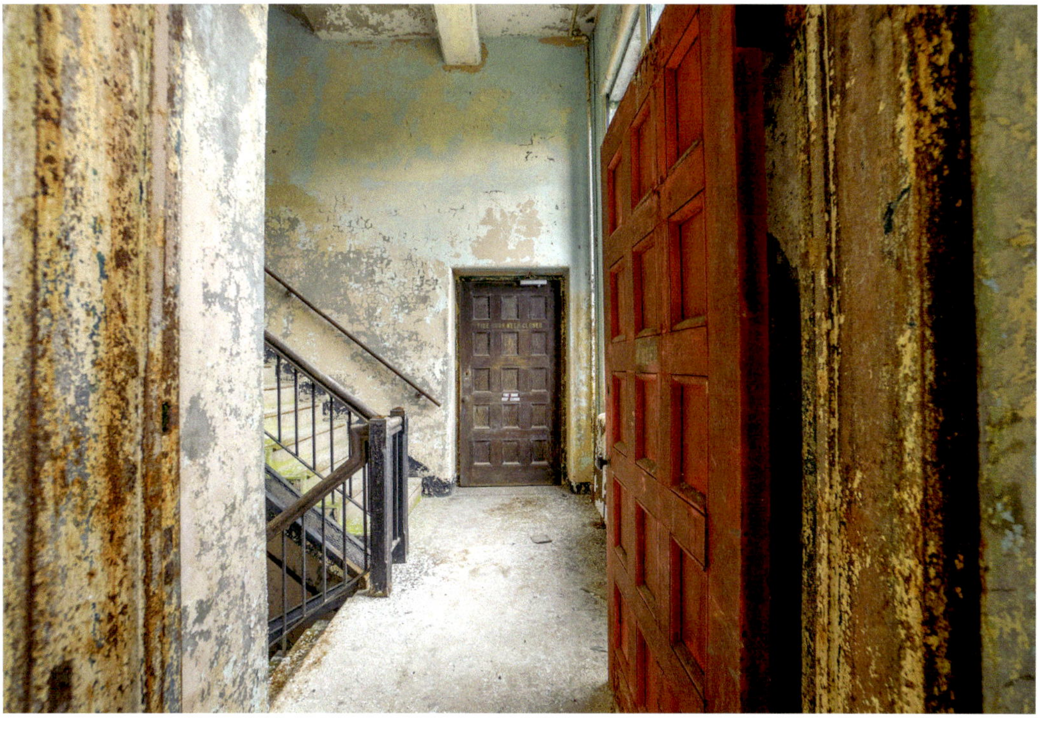

Red fire doors in an asylum originally built as a prison.

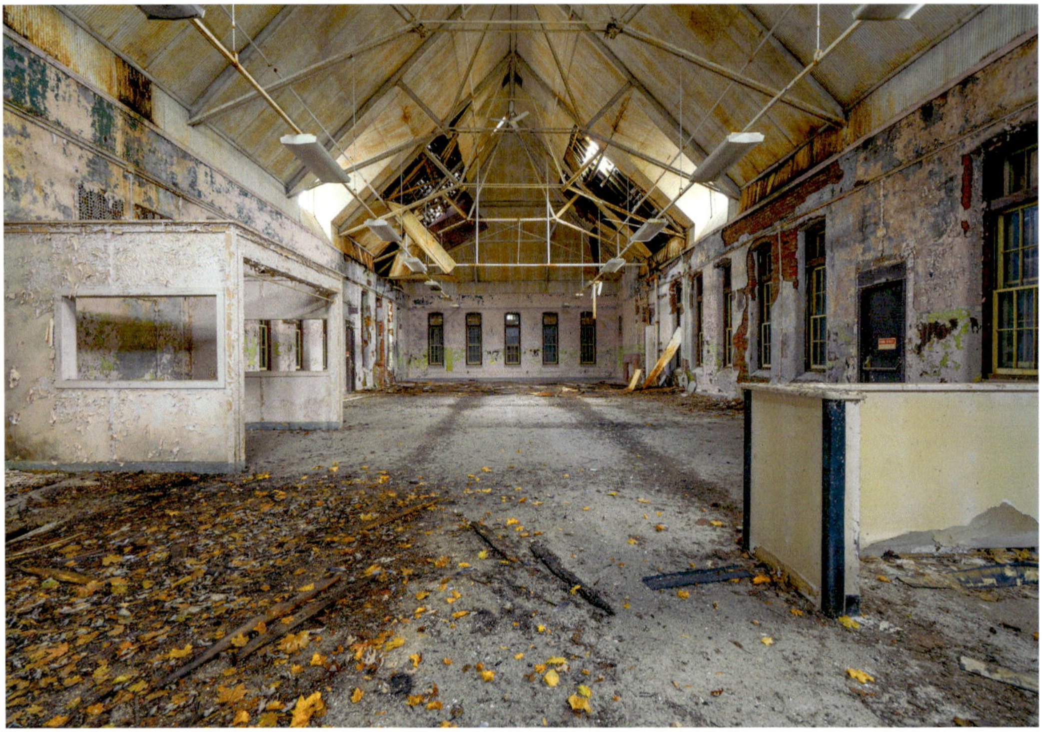

Open wards were beneficial for staff, allowing them to easily monitor larger numbers of patients at a single time.

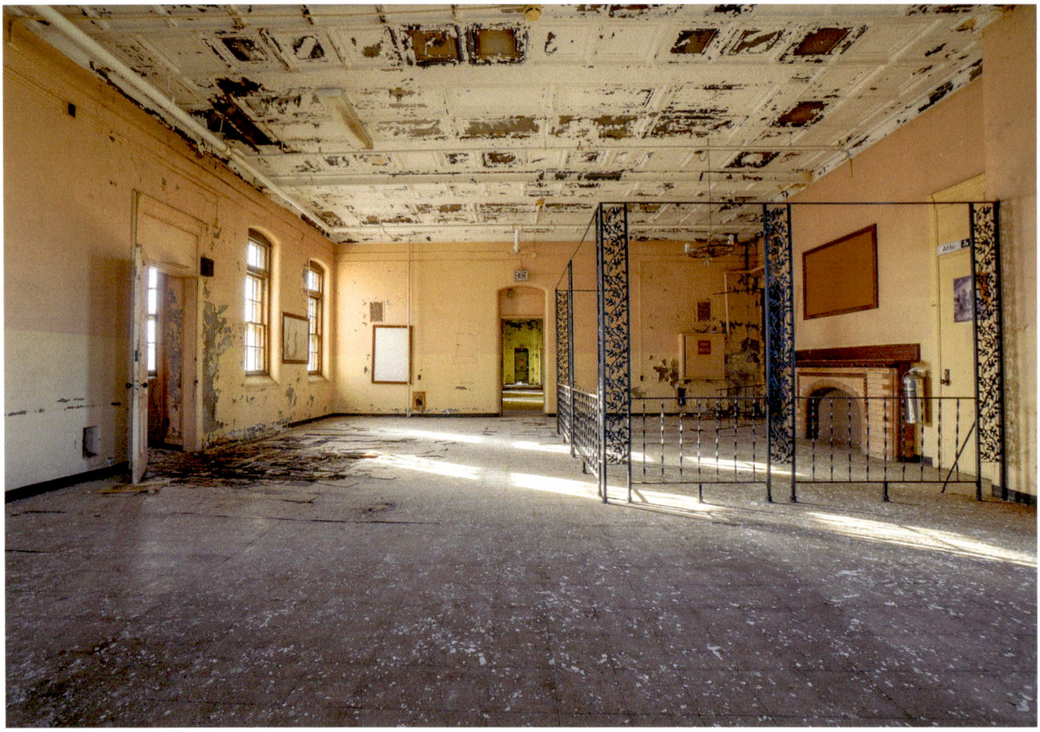

The common areas for the patient were sometimes located around fireplaces or large windows to provide comfort.

# 4

# INFECTIOUS

During the late nineteenth and early twentieth centuries, tuberculosis was a significant disease affecting the United States, particularly the state of New York. This infectious disease caused severe pain in the lungs, hacking coughs, and bloody sputum. In 1882, microbiologist Robert Koch discovered that tuberculosis was caused by bacteria, not genetics, and was highly contagious. By the turn of the century, tuberculosis had killed one in every seven people on the planet.

As the number of tuberculosis cases grew in cities, a new type of institution that could combat the disease became necessary. Sanatoriums were established to treat and "cure" tuberculosis. The only known treatment methods were fresh air, sunlight, exercise, and a healthy diet. American physician Edward Trudeau opened the first sanatorium in Saranac, NY, to help fight the disease that was killing so many New Yorkers. These institutions were built away from overcrowded and polluted cities to enhance the amount of fresh air and sunlight a person receives.

Sanatoriums soon became popular across the U.S., and the first large state and government-funded sanatoriums borrowed elements from Dr. Thomas Kirkbride's theory on airflow, natural light and architectural styles for treatment. These structures had a center administration building, large patient wings on each side, open porches, staff housing, and communal spaces. The added architectural beauty of these buildings helped patients from feeling bored or mundane since they needed long-term treatment.

It was not until the 1940s that Albert Schatz discovered streptomycin, an effective antibiotic for the treatment of tuberculosis. With modern medicines and treatments by the 1950s, the number of sanatorium patients steadily declined, and most, if not all, of these institutions closed before the 1960s.

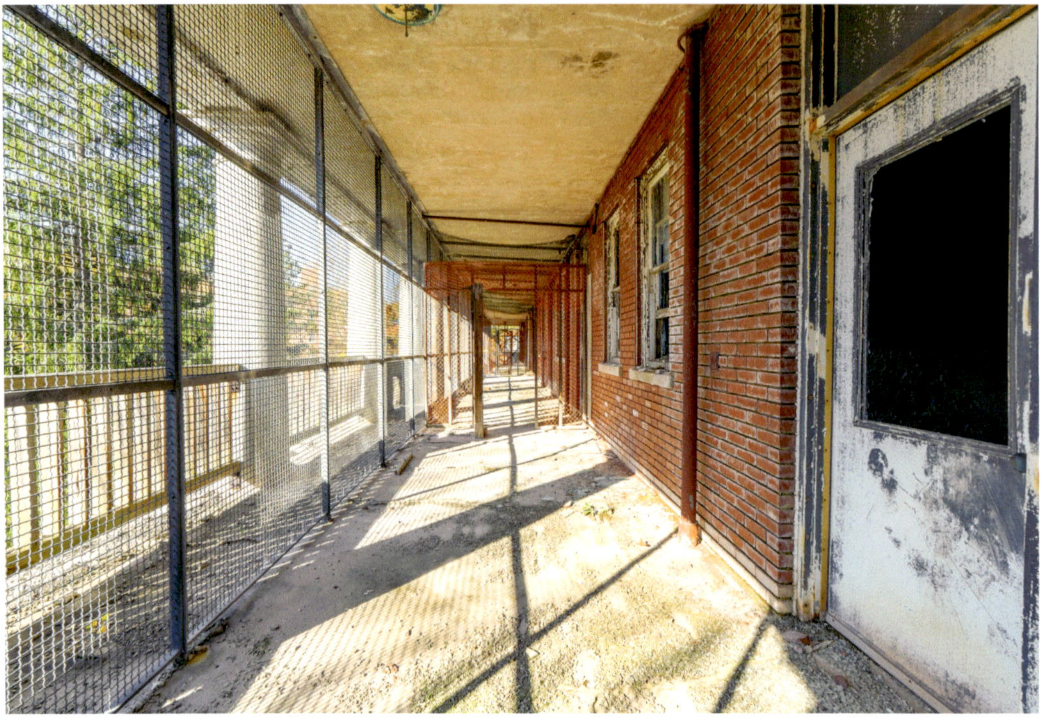

Large open porches were necessary in the early fight to cure tuberculosis.

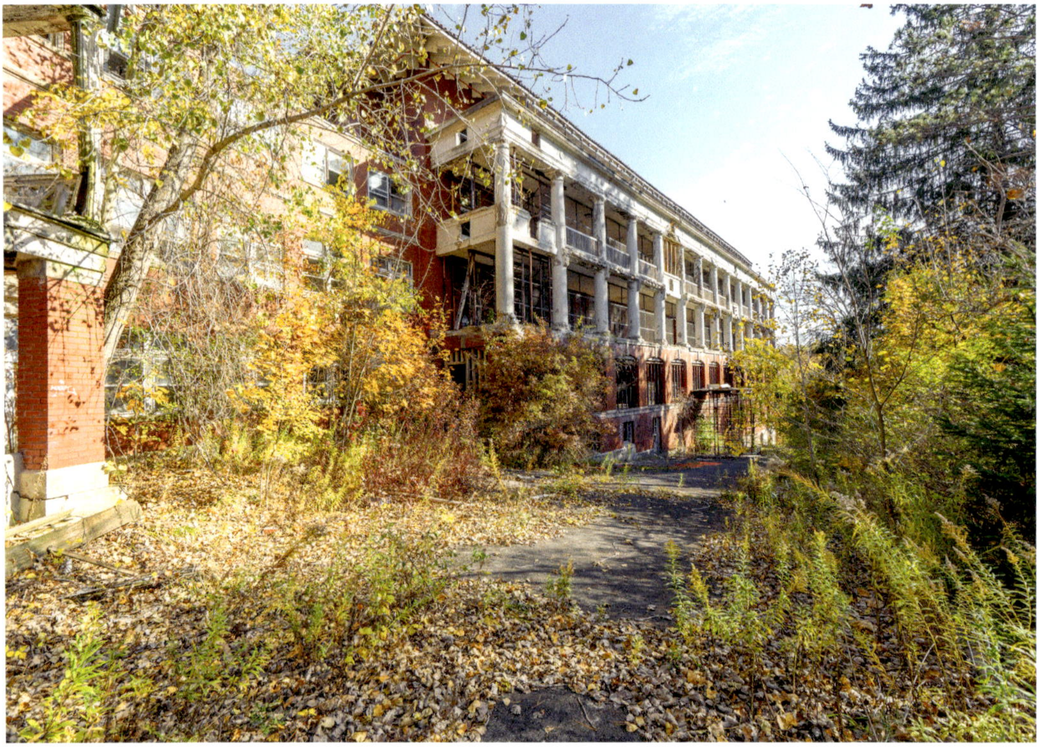

Every patient room/ward must be accessible to a porch or the outside.

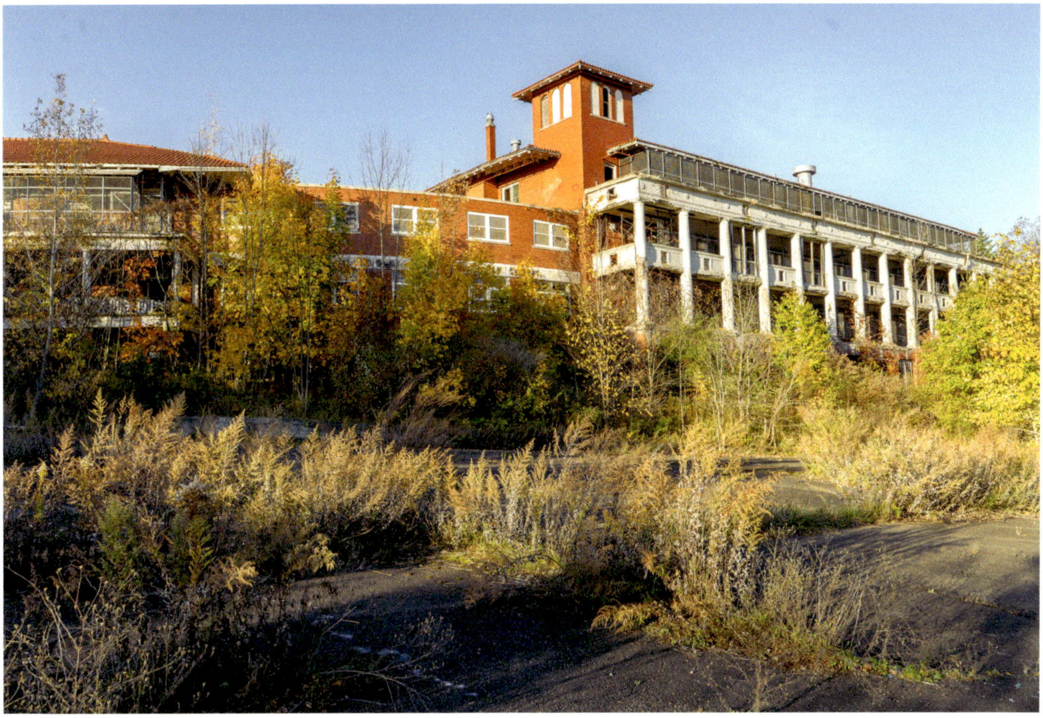

Tuberculosis hospitals were set in the scenic countryside away from populated and polluted cities.

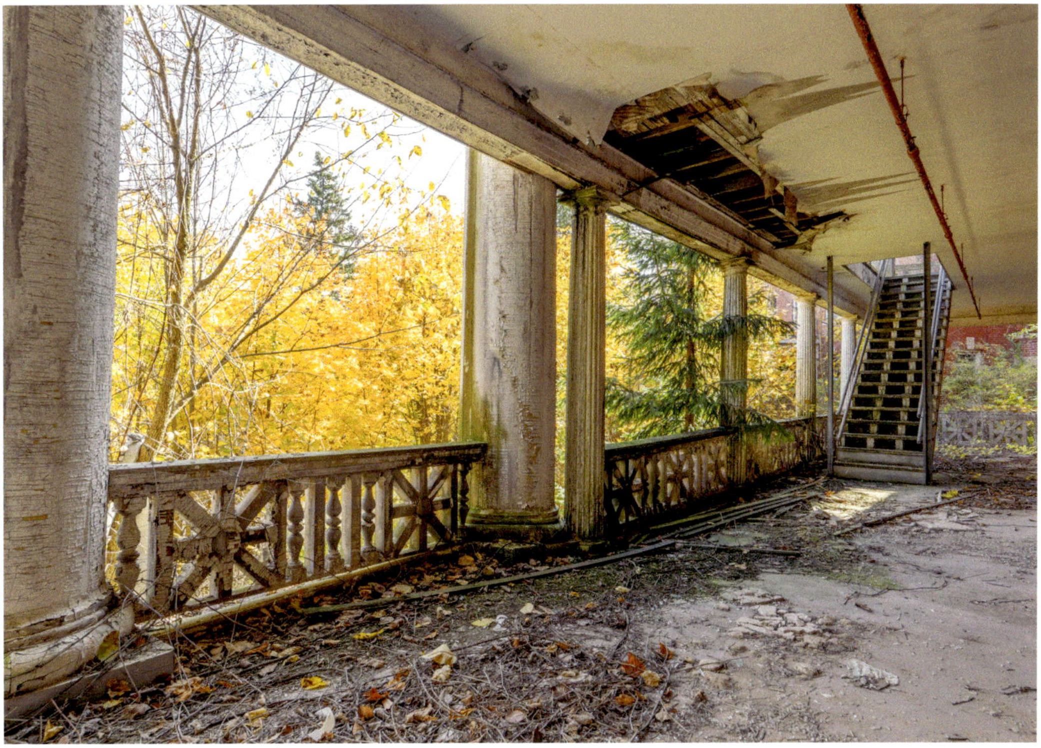

This TB hospital was modeled after southern plantations with ornamental columns and wide verandahs.

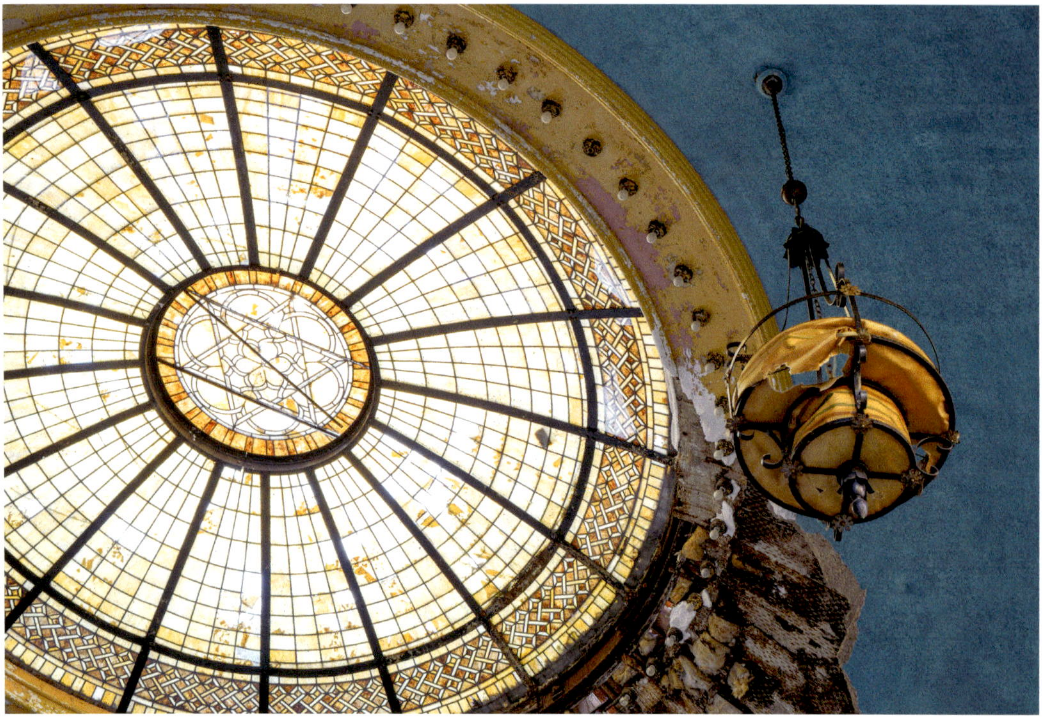

This glass dome wasn't just for beauty but also for heliotherapy.

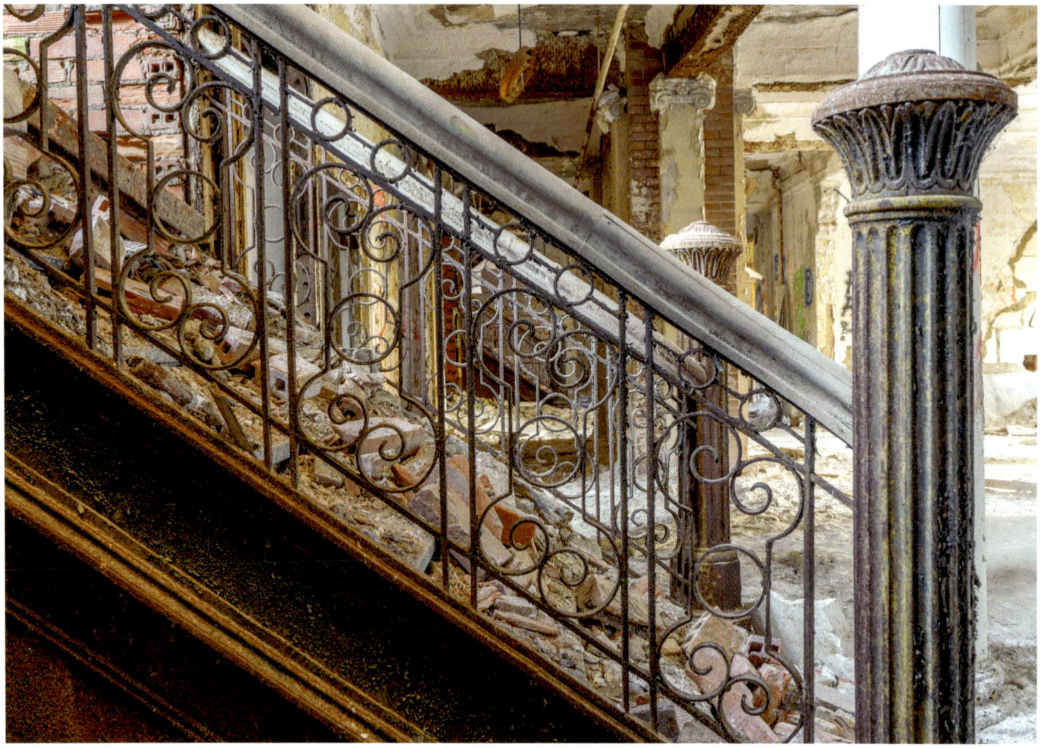

One of the two decorative main staircases in the administration building that would greet patients and guests.

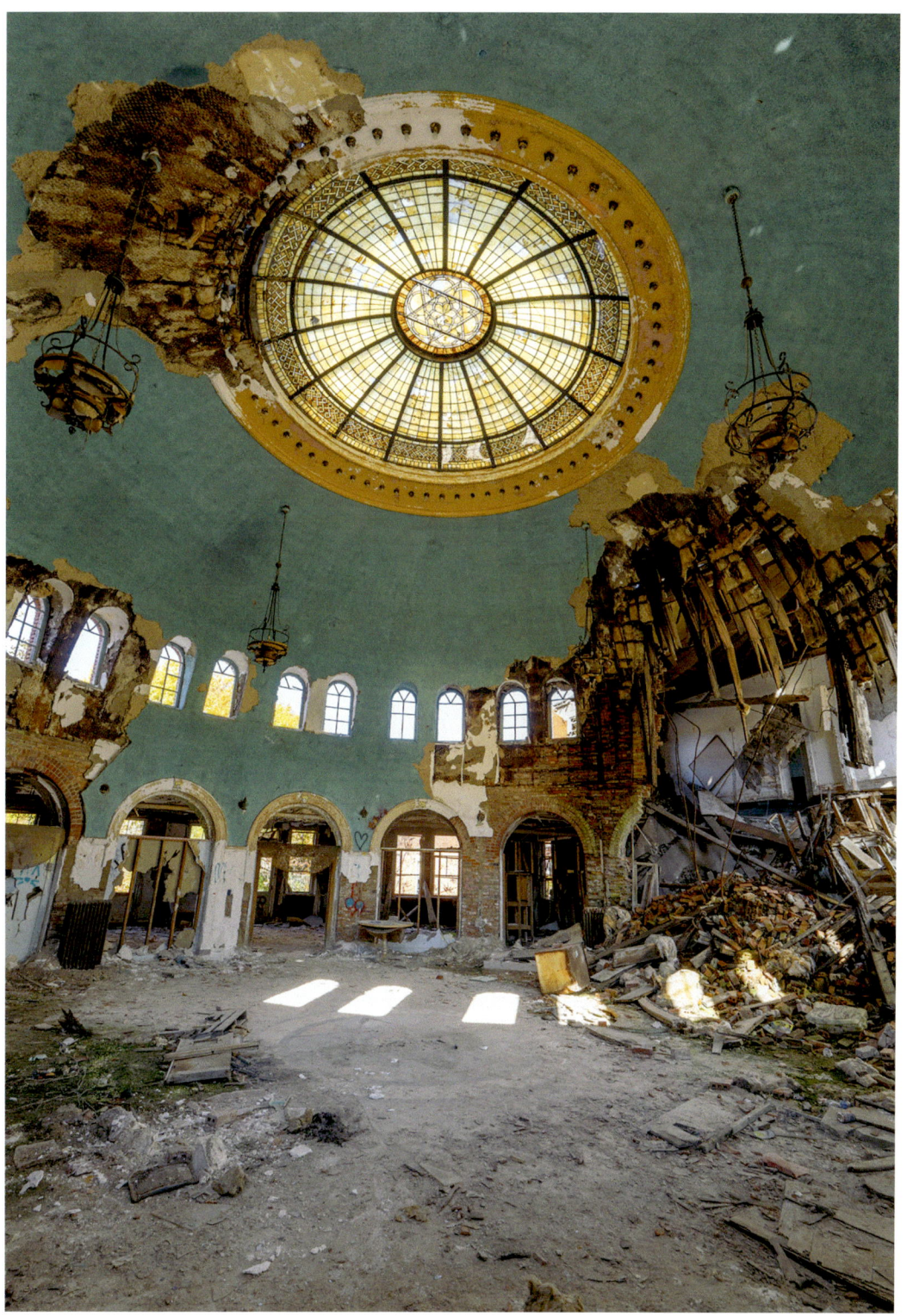
The patient dining hall dome was designed to bring as much natural light as possible and to look like a sunny summer day.

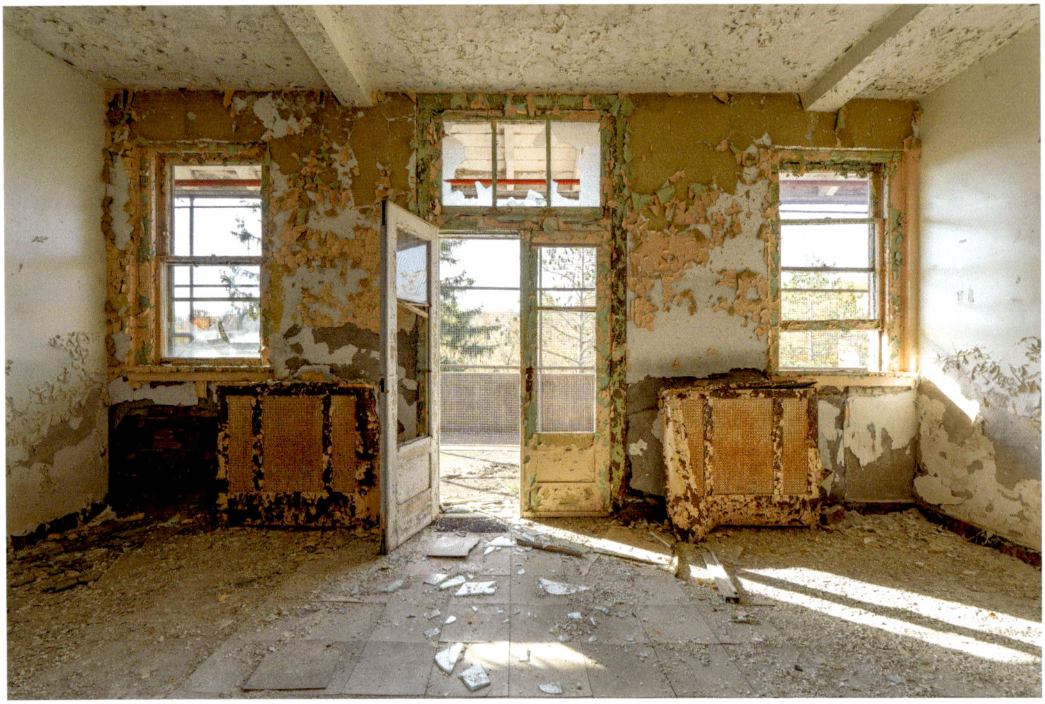

Private rooms were not always available, and many patients had to share space with others who were also ill.

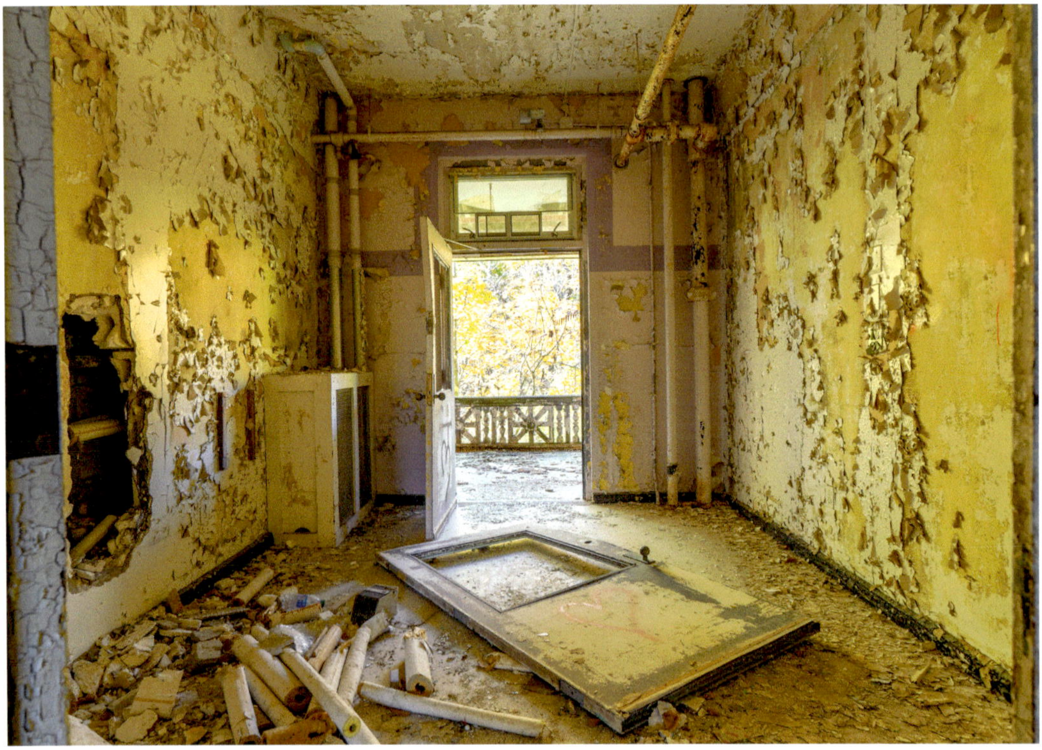

A private patient room opening up to the sun porches.

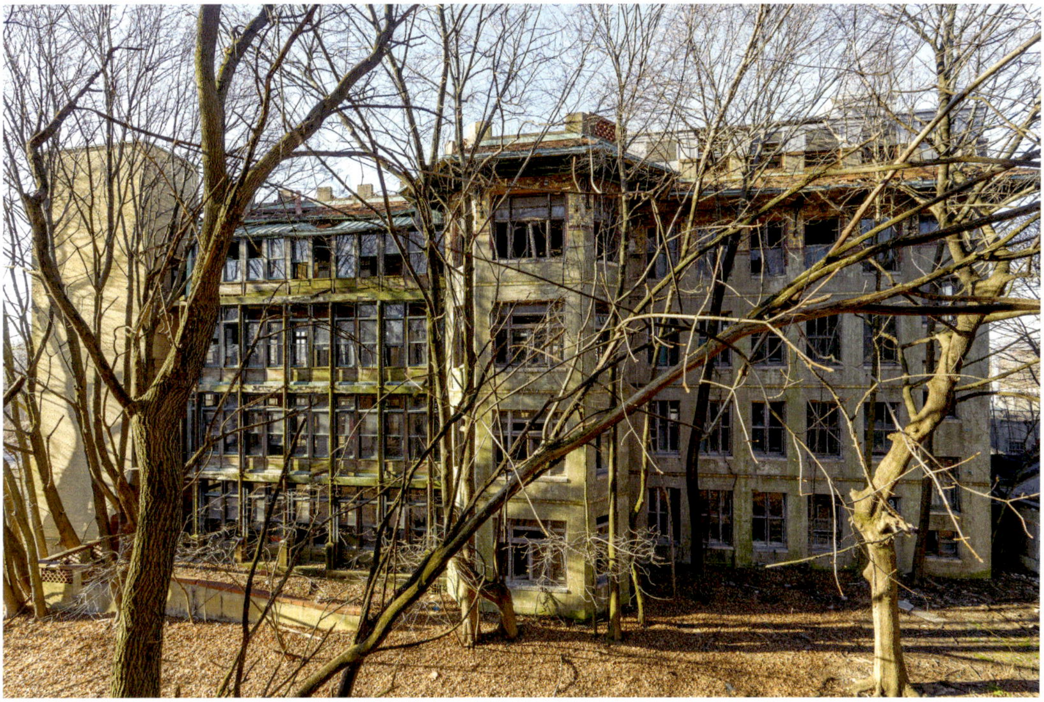

This tuberculosis hospital was designed to make patients feel like they were on a warm vacation destination with fresh air and views of the Atlantic Ocean, rather than fighting a deadly disease.

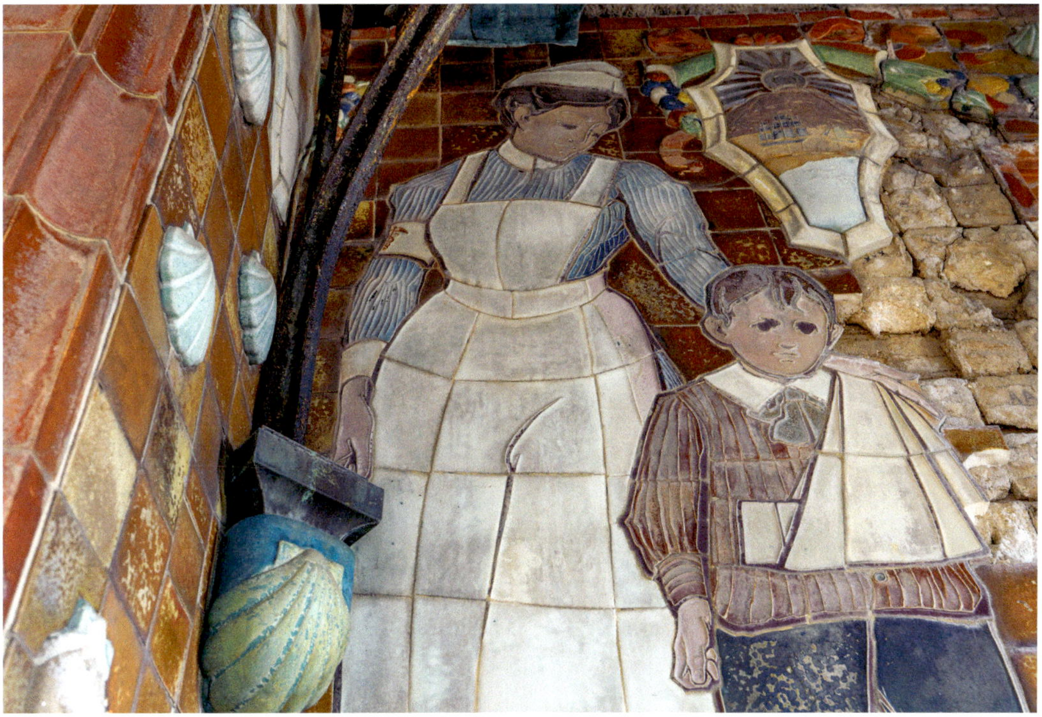

The largest tuberculosis hospital in New York had beautifully detailed tiles on the top floors of each pavilion, including one showing a nurse assisting a young boy.

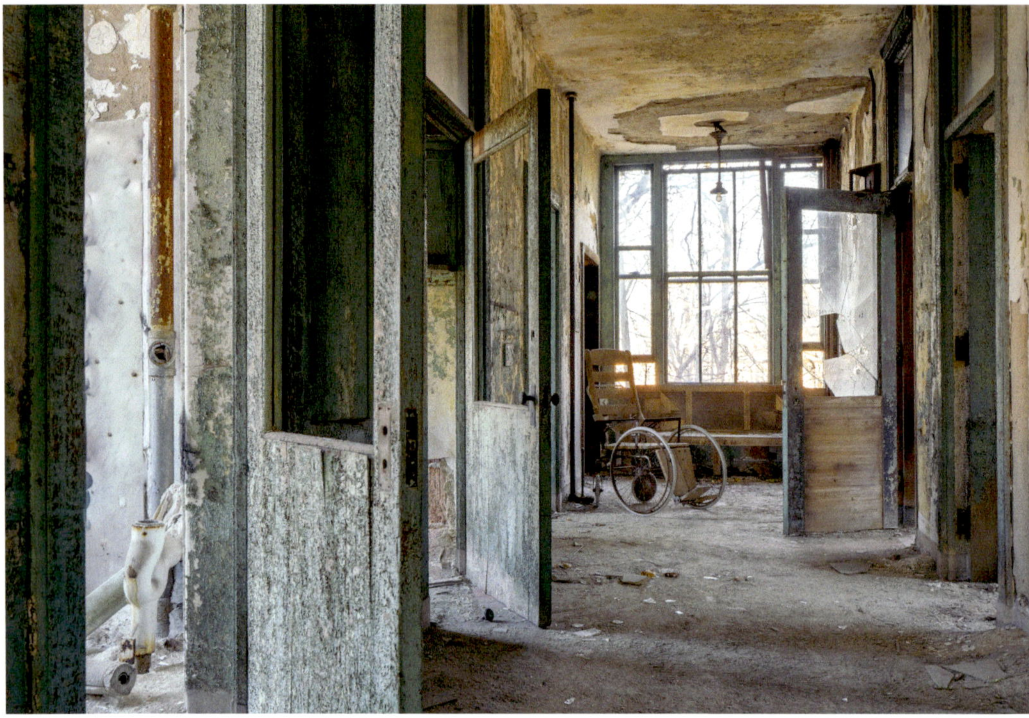

Hallway of the largest tuberculosis hospital in New York, which housed over 2,000 patients and features the first maternity ward in a tuberculosis hospital.

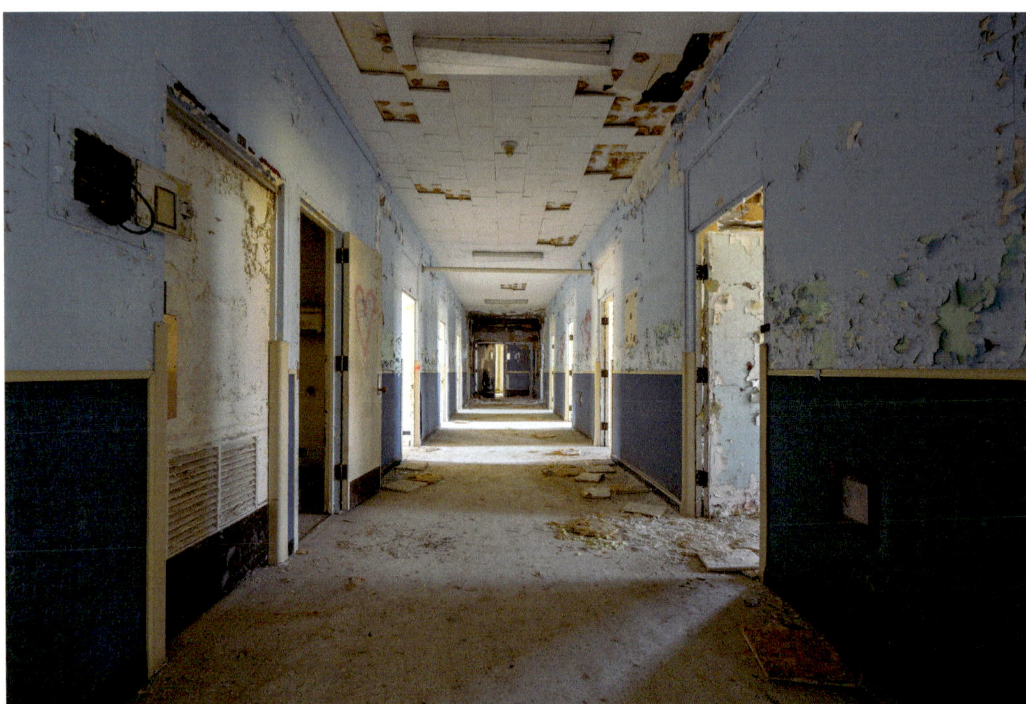

Renovated hallway of a former TB hospital that was later converted into a developmental center before closing.

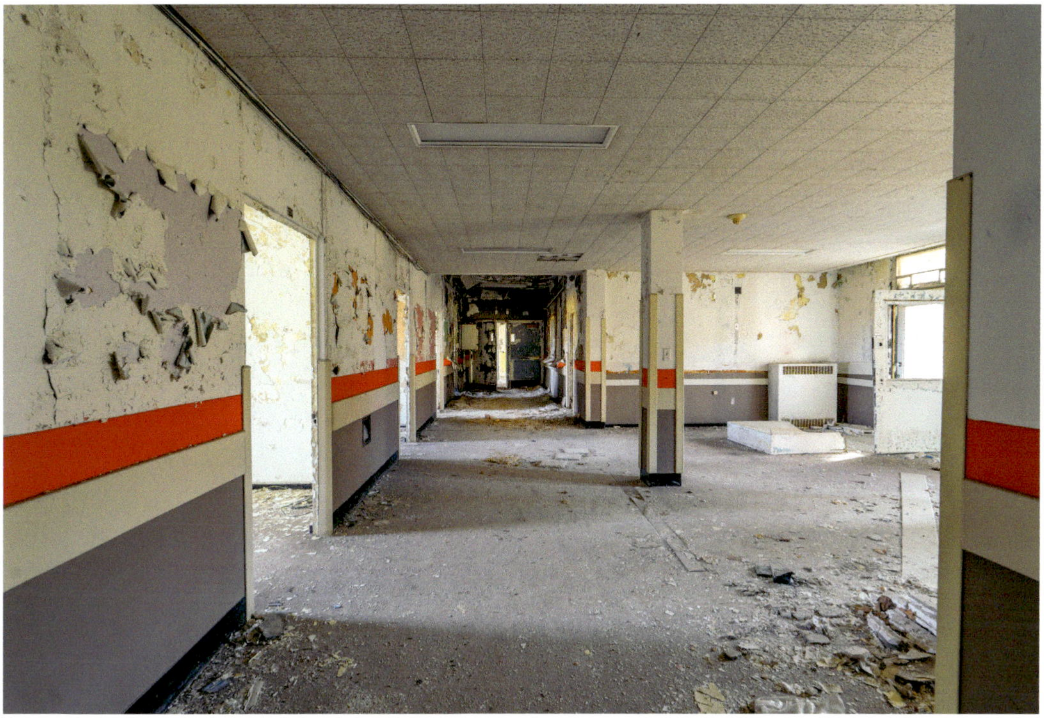

Hallway and dayroom that led to sun porches.

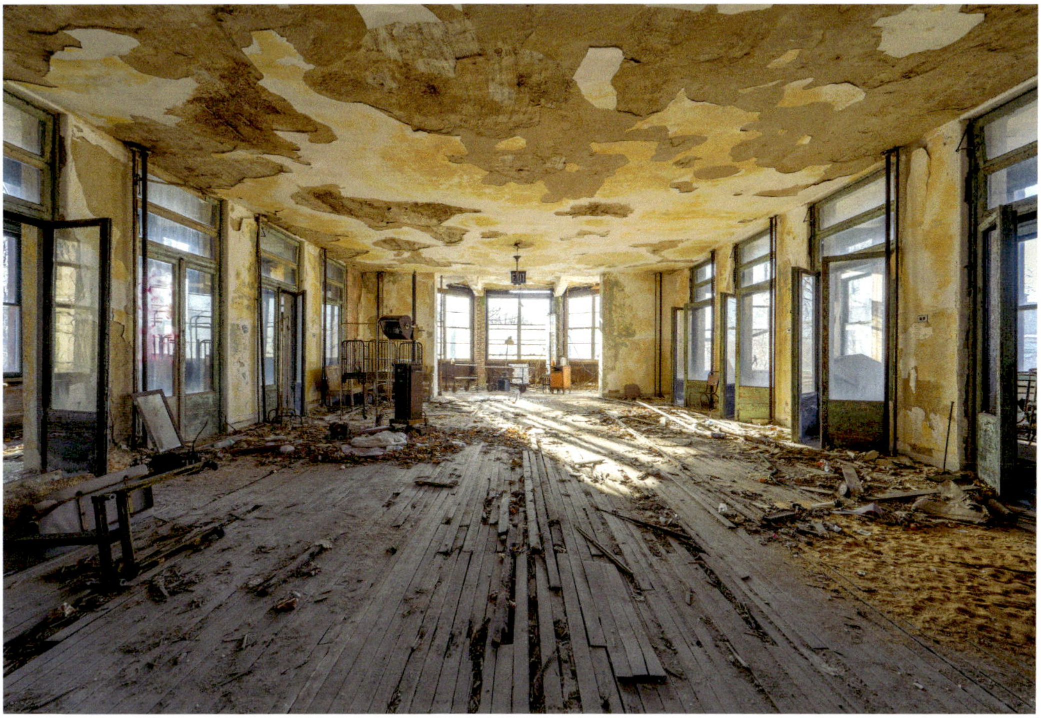

New York's largest TB hospital had open wards and sun porches on each side to allow for ample air and sunlight.

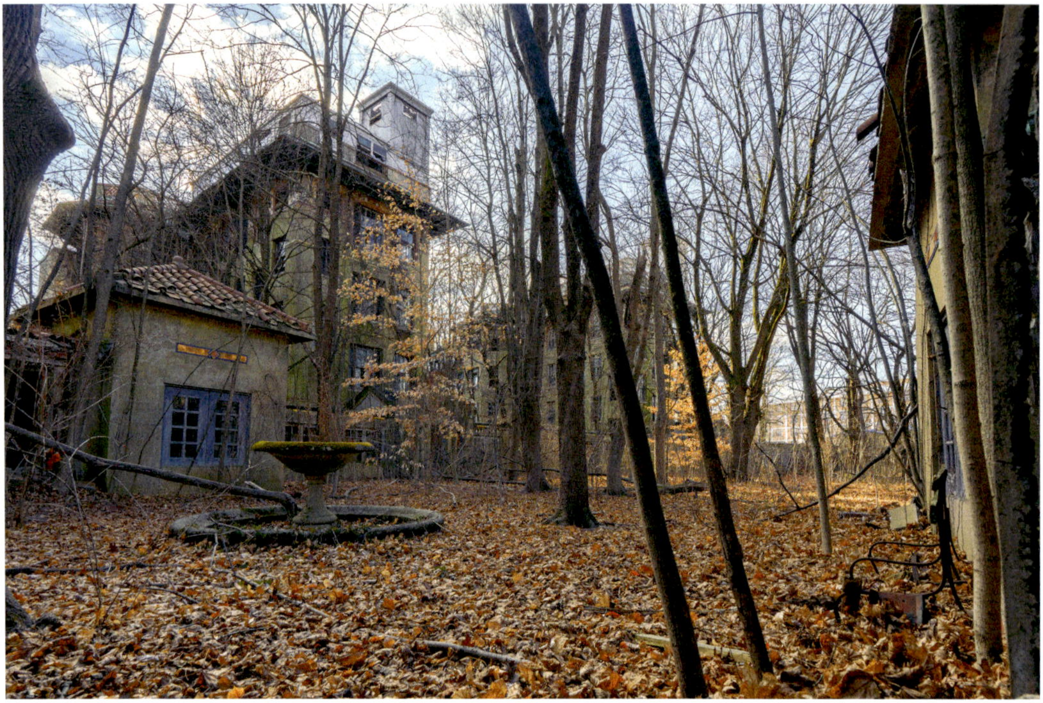

This once beautiful courtyard features a fountain, greenery, and breezeways connecting the dining hall to the female pavilions.

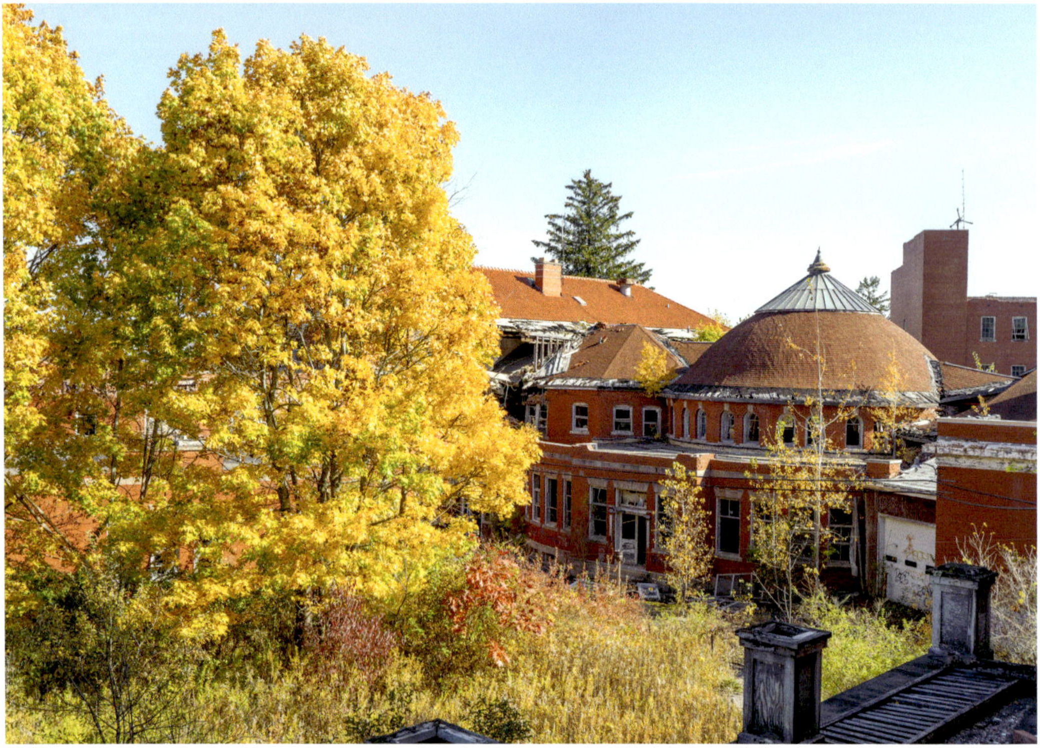

View from the staff dorms of the east side of the administration, dining hall, medical building and theater.

# 5

# RECREATION

In mental health institutions, recreational activities play an essential role in the treatment of patients. These activities helped improve the overall health of their mind by stimulating their brain functions and promoting leadership skills, proper social behavior, judgment, and emotional control. The hospital administration considered it crucial to the patient's recovery process. Some state hospitals even had state-run sports teams like basketball, baseball, and bowling that would compete against other state hospitals or local towns in tournaments. Apart from physical activities, patients were encouraged to participate in the arts, including live theater, to express themselves as well. Social events, such as holiday parties and dances, were also organized for patients to enjoy with their loved ones. The institutions offered various classes and activities such as crafts, music, recreational, and industrial therapy to improve patients' treatment and overall mental health.

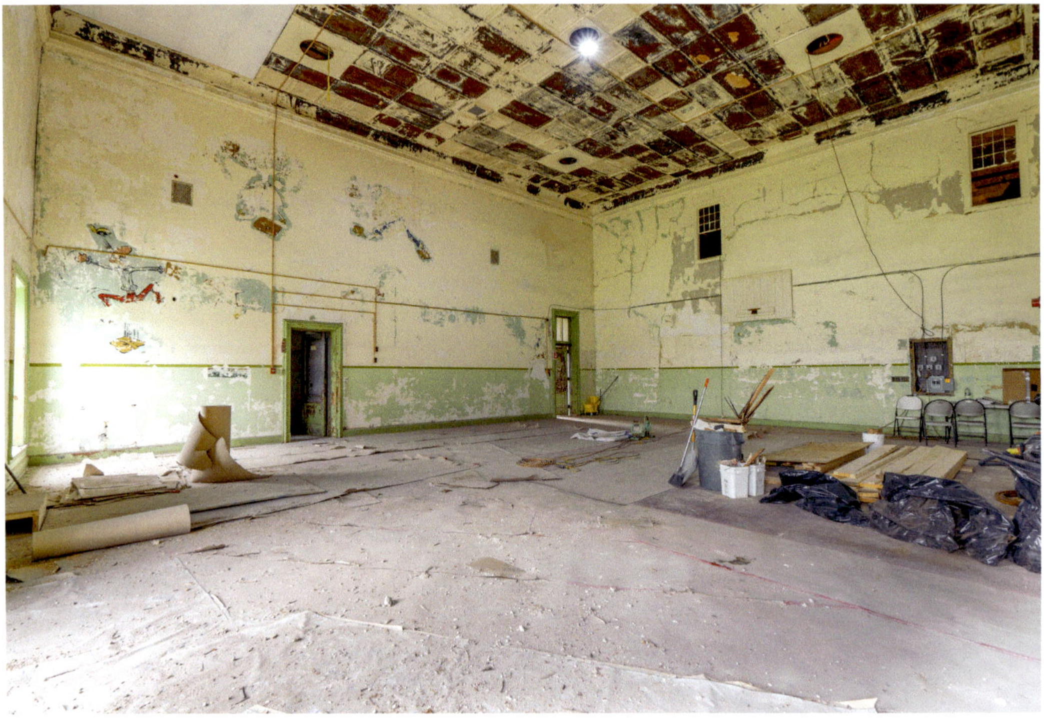

Gymnasiums were built to allow patients to play popular sports like gymnastics, tennis, and basketball.

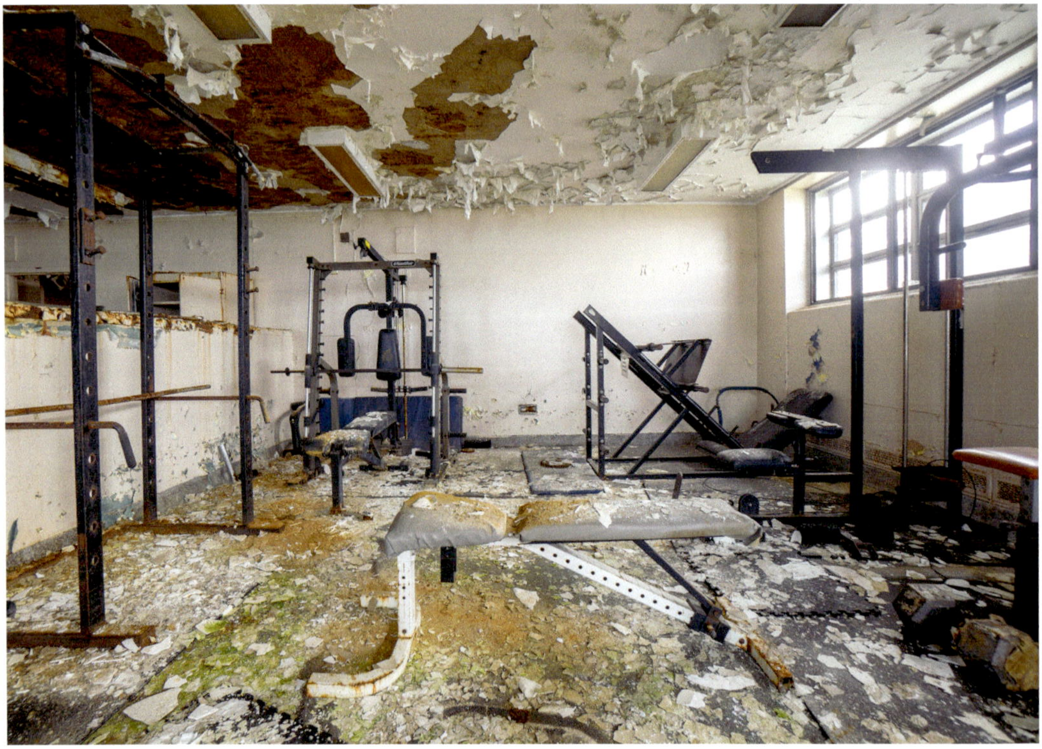

Exercising was essential for overall mental and physical health.

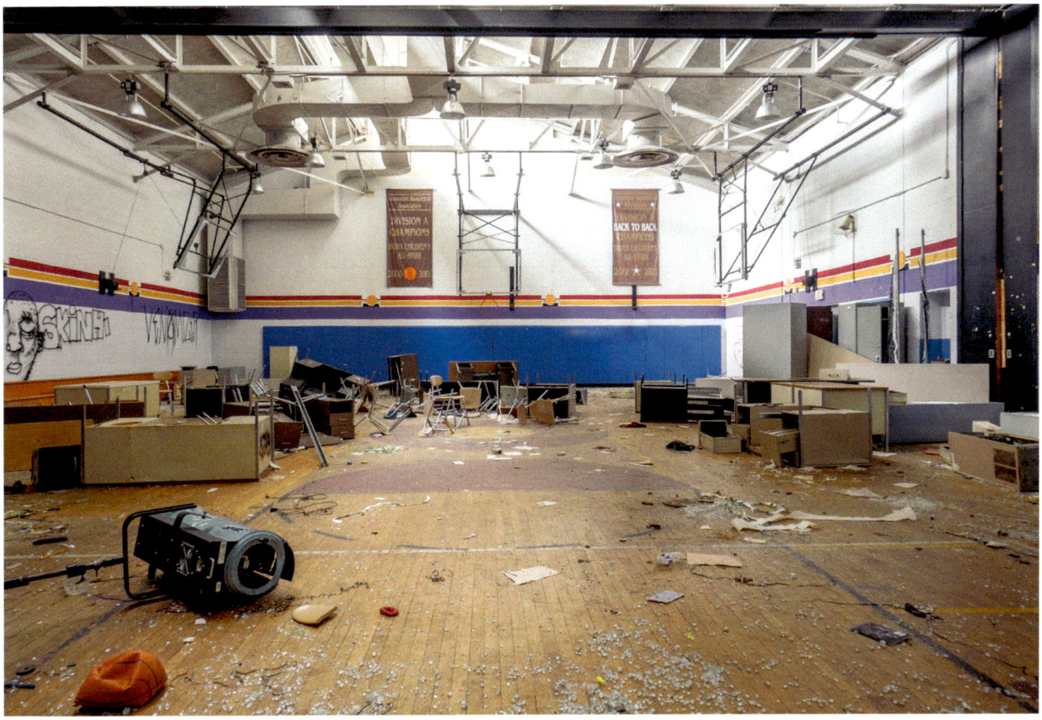

Competing in sports would help patients with purpose and drive.

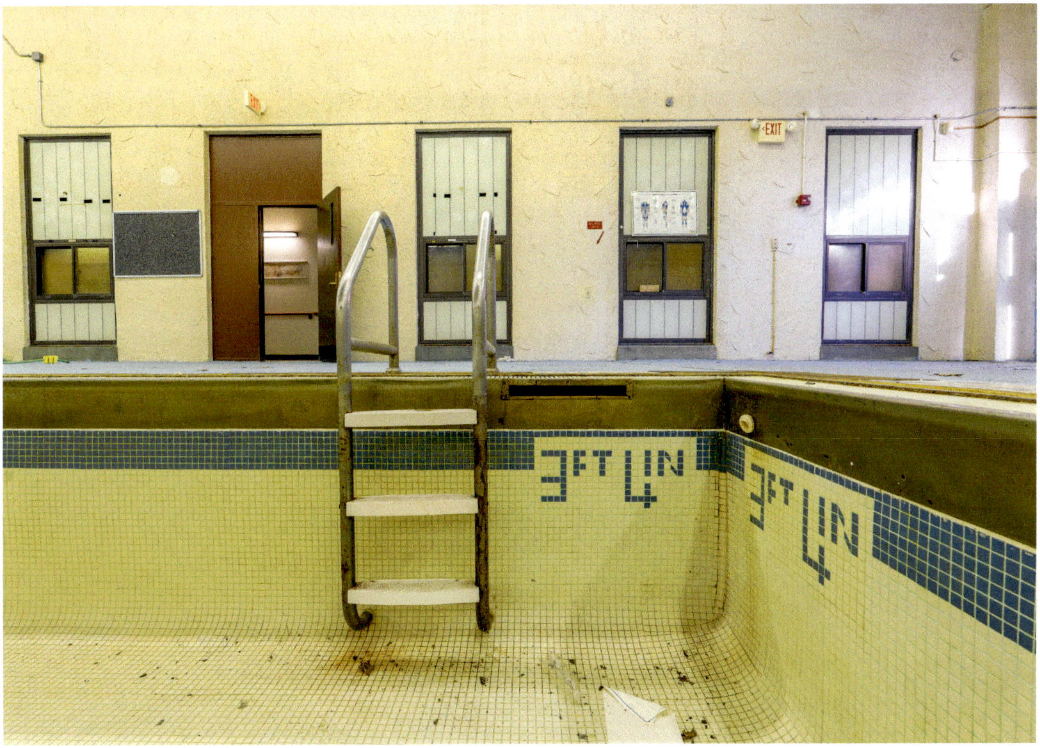

Some institutions offer swimming as a recreational and therapeutic activity.

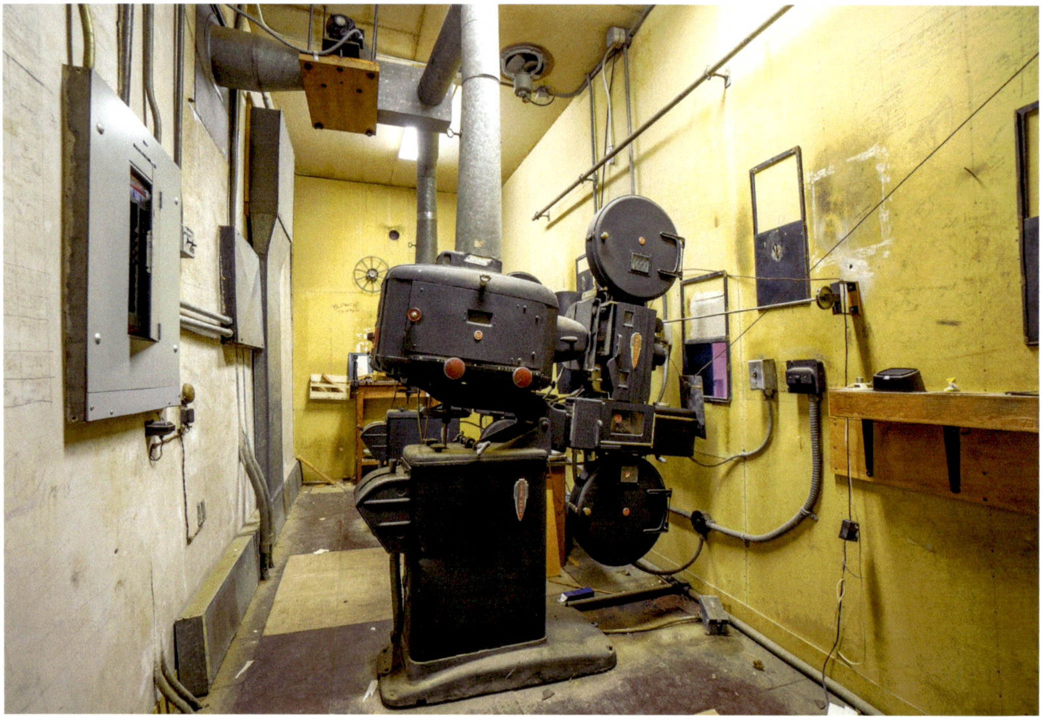

Movies were a great way for patients to be entertained and to escape asylum life.

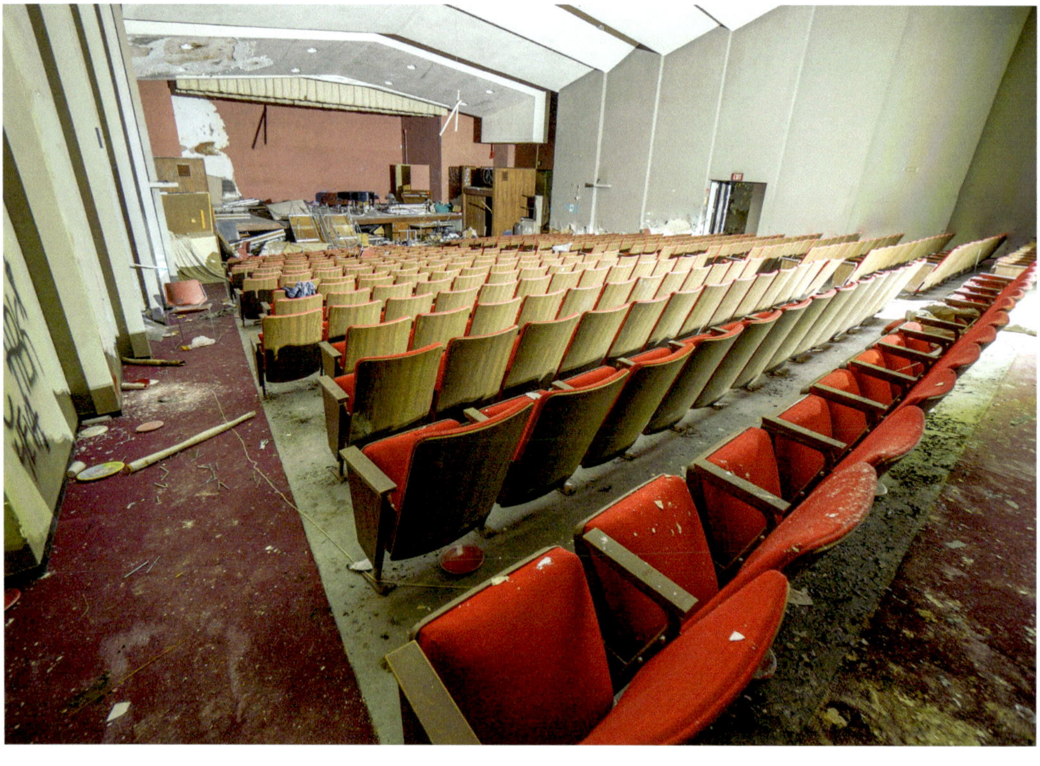

Contemporary theaters were designed to prioritize accessibility over aesthetics.

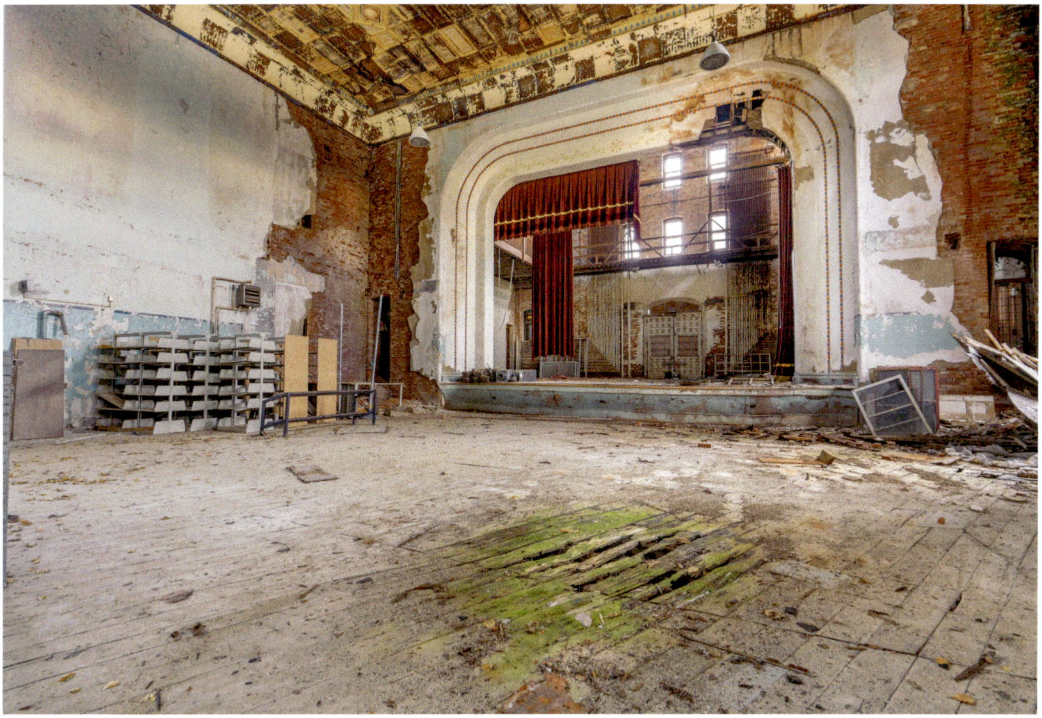

This old theater has been converted into a recreation center but retains some of its original beauty.

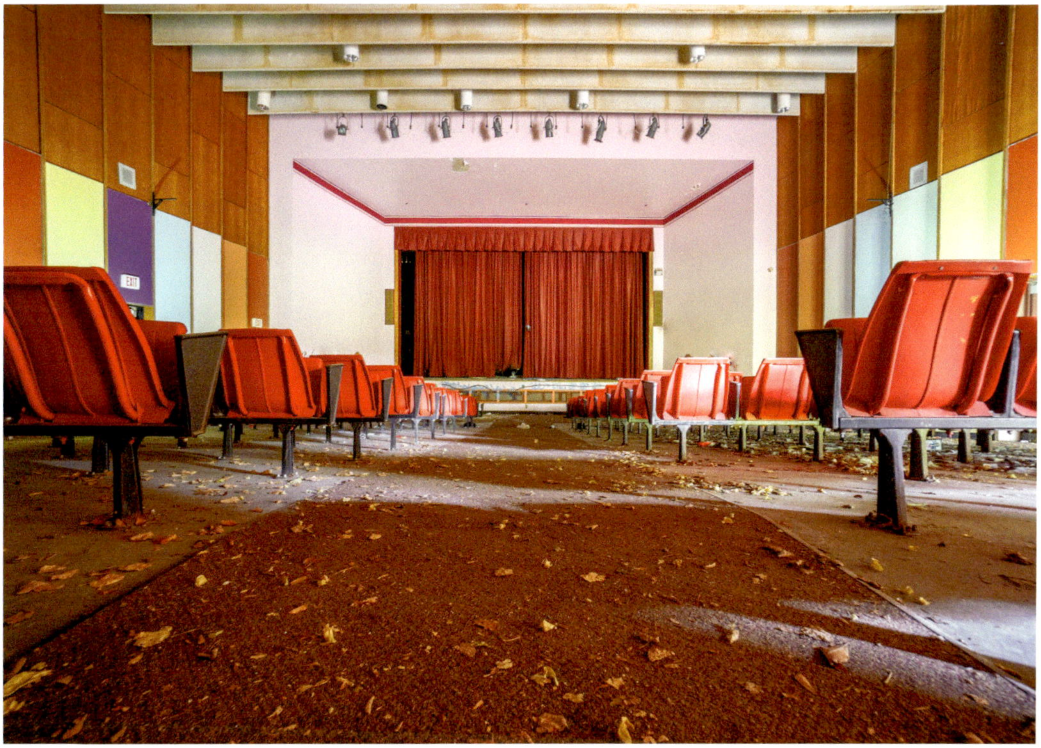

Theaters were also used for patient performances, dances, and parties.

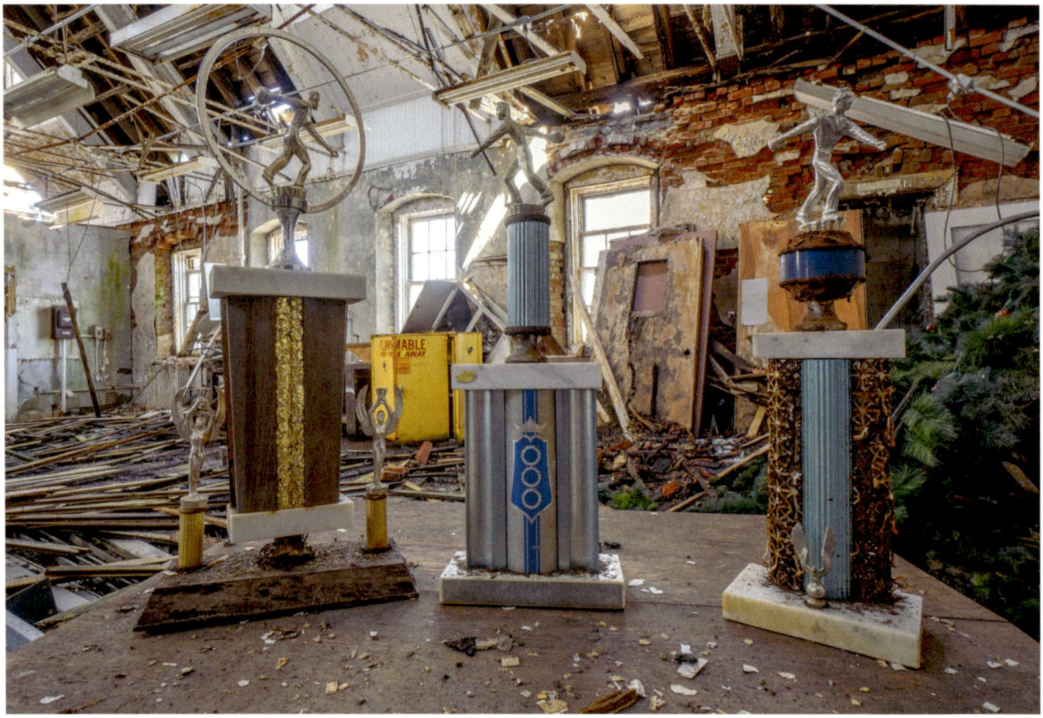

Patients and staff would compete with other hospitals or local towns for championship trophies.

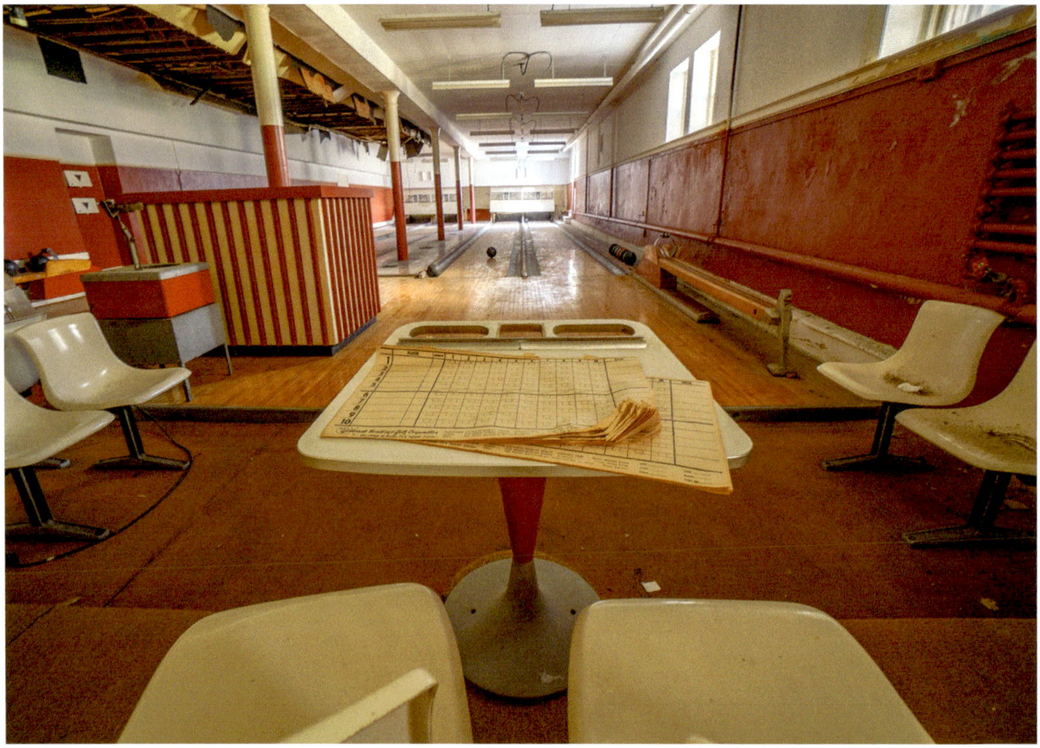

Most recreation center's bowling alleys were located in the basement under the theater or gymnasium.

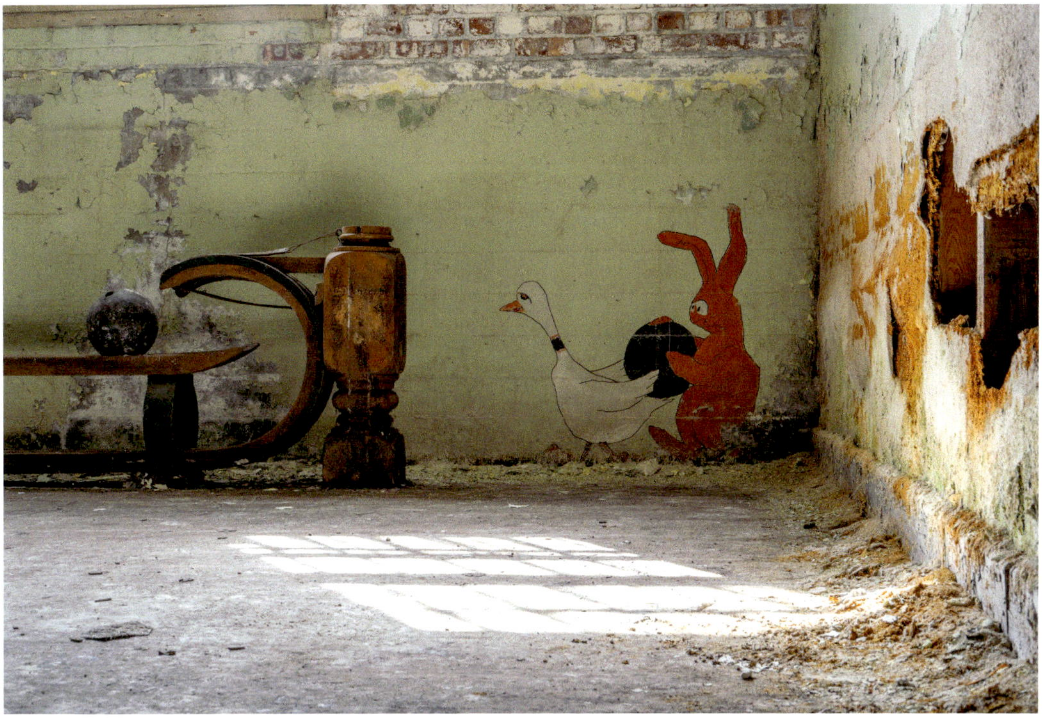

In the early days of asylums, bowling was a popular activity that was only offered to male patients and staff.

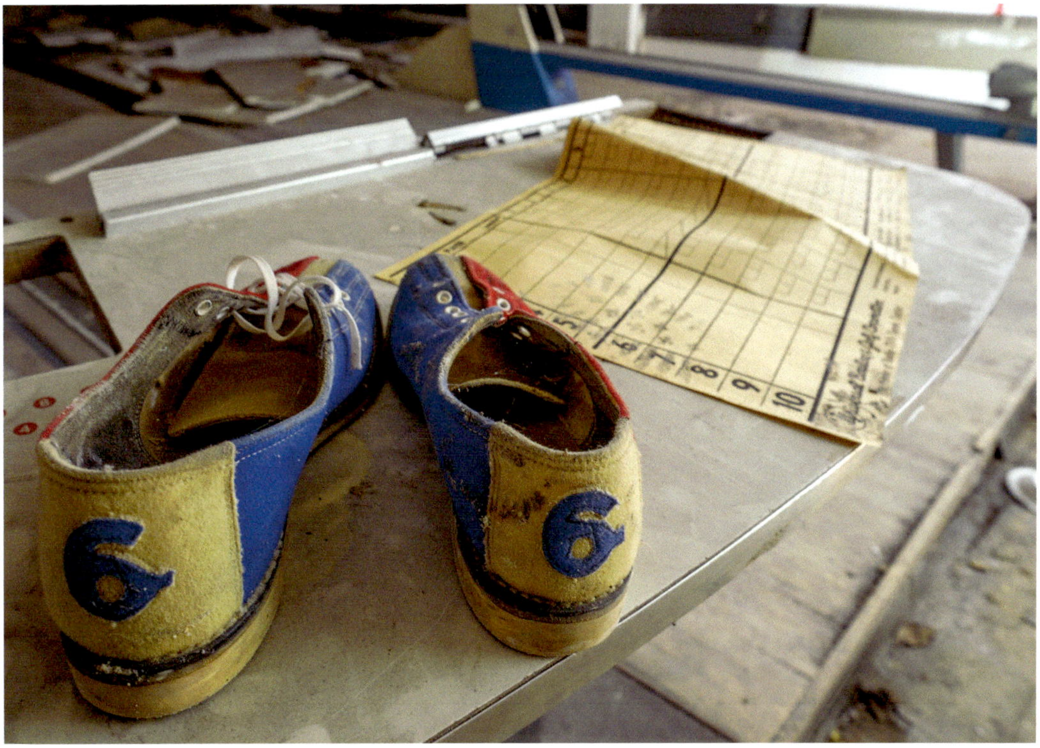

A patient's bowling shoes sit next to an abandoned blank scorecard.

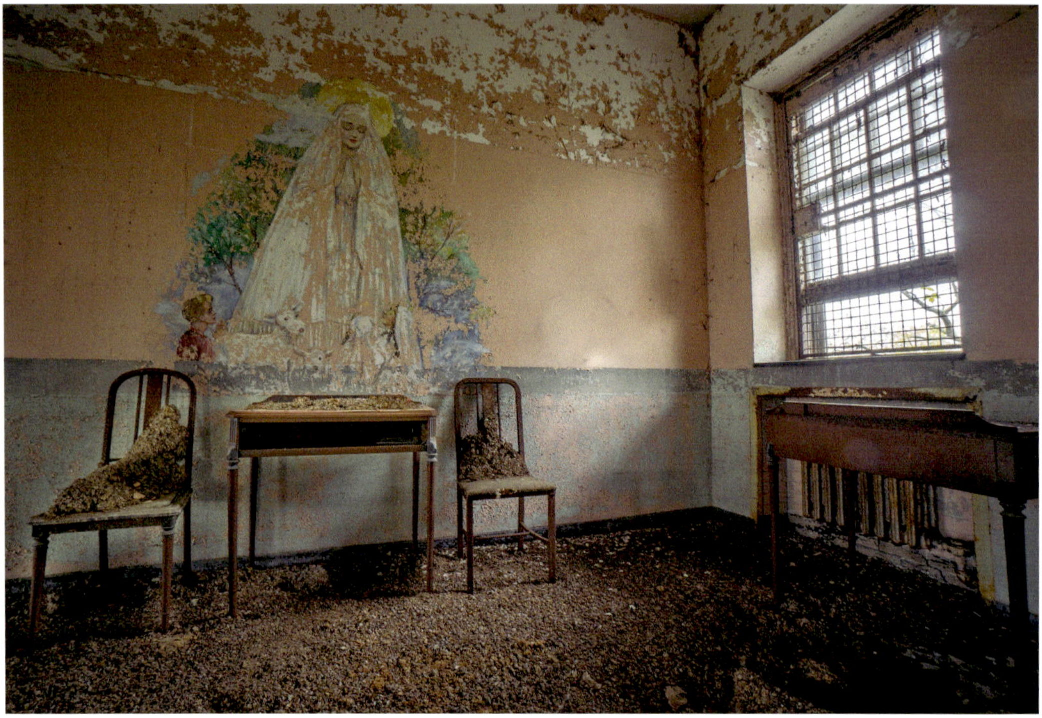

Art is an excellent medium for self-expression in recovery.

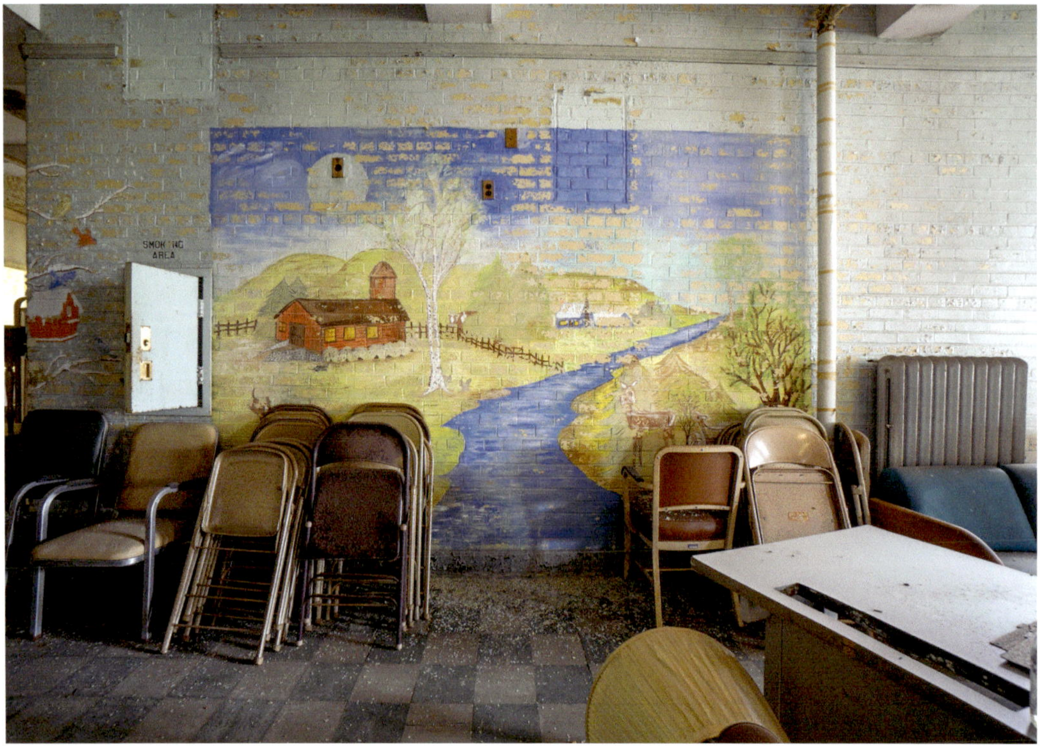

Patients often create paintings of scenes to decorate the dayroom and halls.

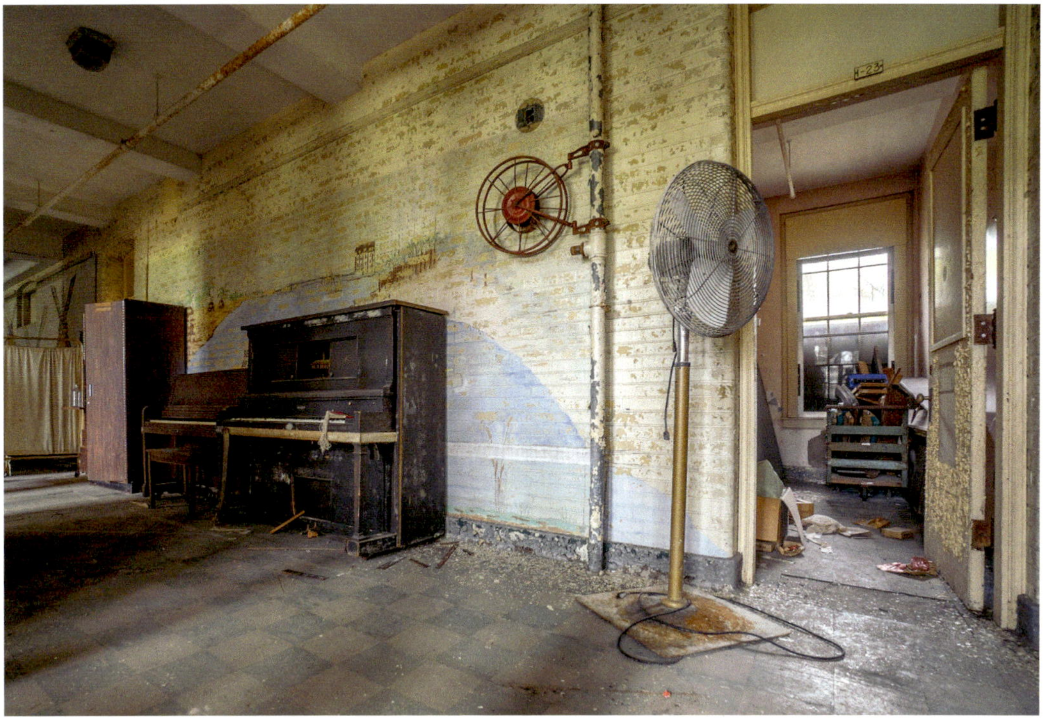

Music was one of the most popular activities patients would enjoy.

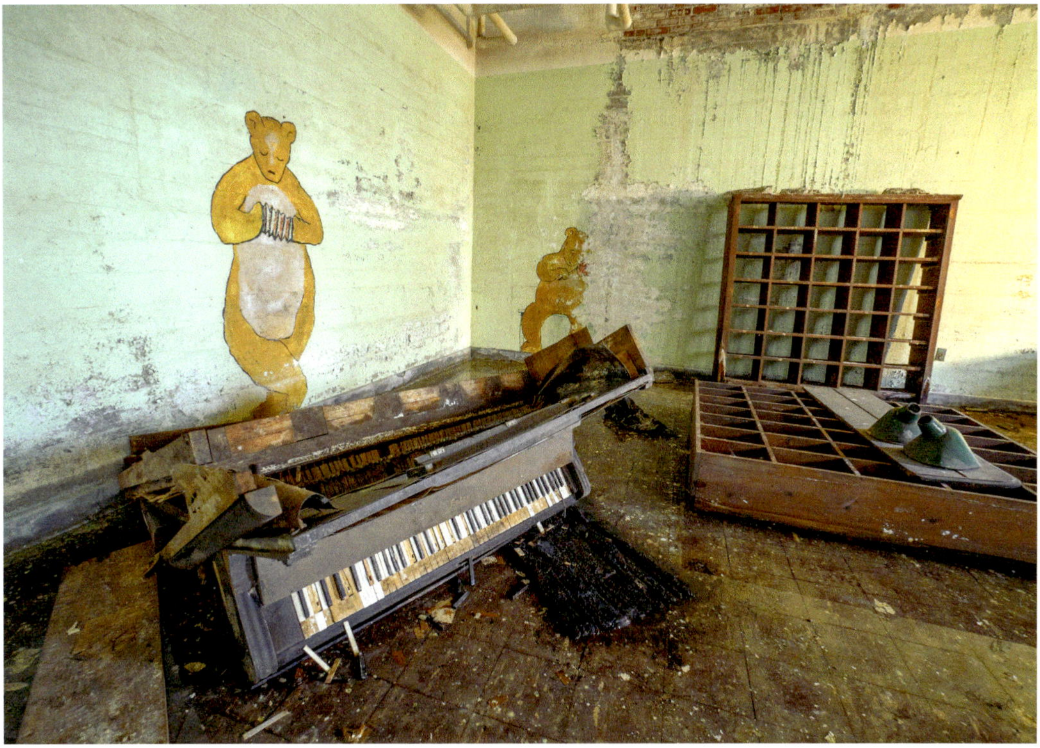

Music was thought to be so beneficial that doctors would encourage it.

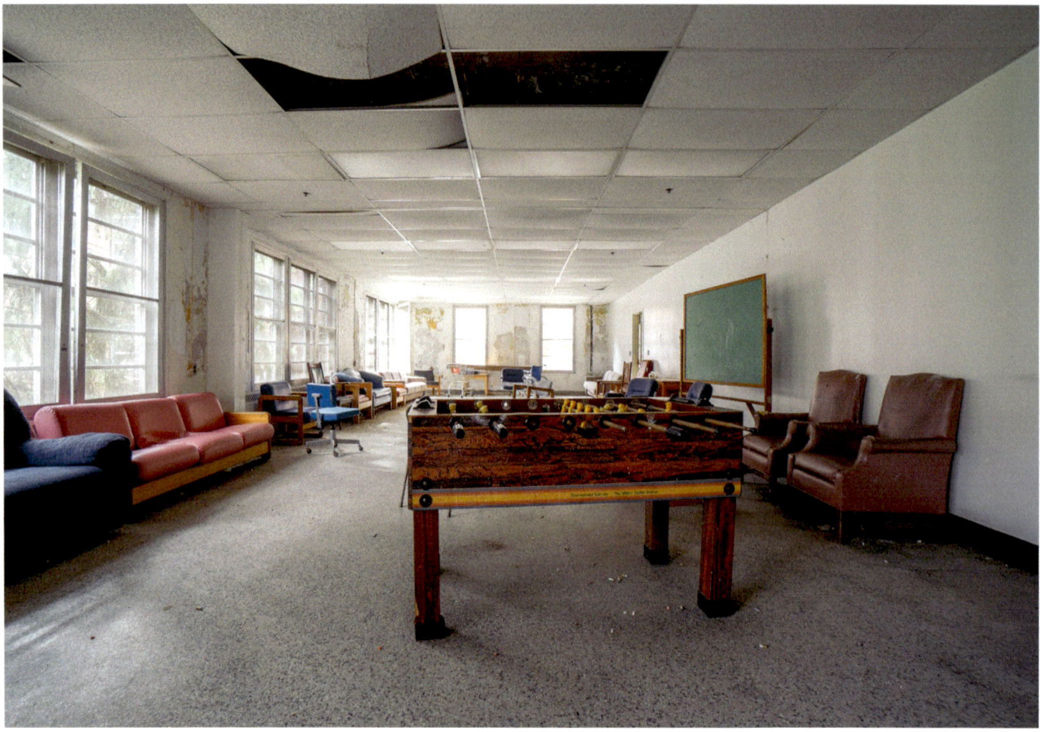

A friendly game of foosball can improve patient mood and encourage social interactions.

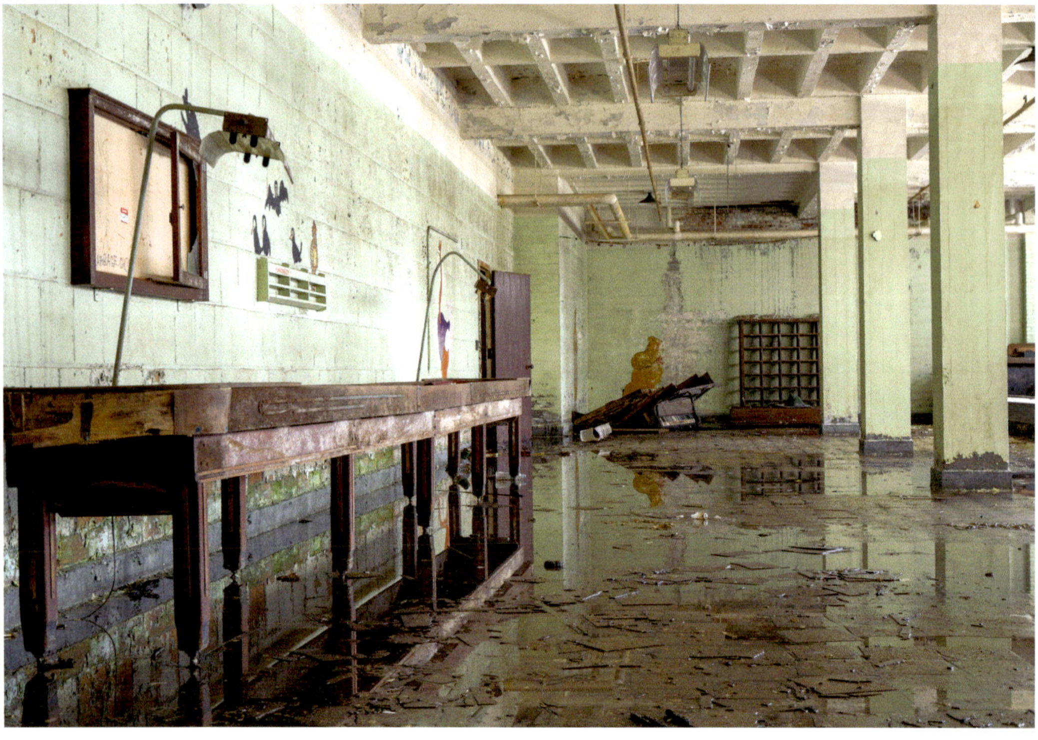

This activity center included a piano, shuffleboard, and a two-lane bowling alley.

# 6

# EUGENICS

Eugenics was a prevalent but very forgotten movement in the United States. Although the United States did not invent eugenics, it gained popularity globally in the 1920s, particularly after World War I, when the U.S. became a world power and inspired other countries to follow its lead. To comprehend the impact this movement had on the state of New York, it is essential first to understand what eugenics is. It is "the study of or belief in the possibility of improving the qualities of the human species or a human population, especially by discouraging reproduction by people presumed to have inheritable undesirable traits." This would mean the need to spectate certain types of people based on their race and developmental disabilities and with proponents of involuntary sterilization for racial improvements. The children and adults of the undesirable, who were often already in institutions, would be most affected by this movement and vulnerable to the laws of the land. This movement was so powerful that Adolf Hitler credited the American eugenics movement in his book *Mein Kampf*.

This movement would have a significant effect on the state of New York. In the late 1900s, with the growing population of feeble-minded people and the eugenics movement, they needed an institution that would serve this objective. To carry this out, they sought advice from Dr. Walter Fernald, a prominent superintendent in Massachusetts whose institution for the feeble-minded was considered a model for educating individuals with mental disabilities. He advised on the design, layout, and purpose of a new, self-contained, and self-sufficient institution. His views were taken into account, and instead of large, towering buildings, smaller one-story cottages were chosen to be built around the campus. This decision would create a more humane experience and a village-like atmosphere.

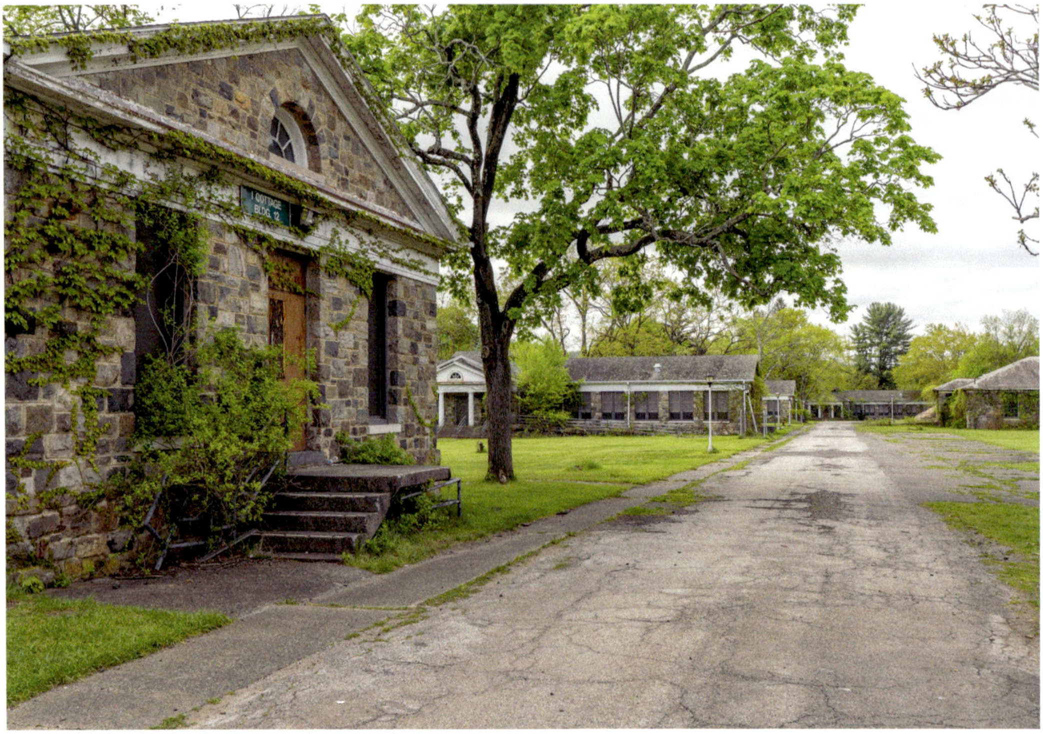

This institution for the feeble-minded classified individuals into three categories: idiot, imbecile, and moron.

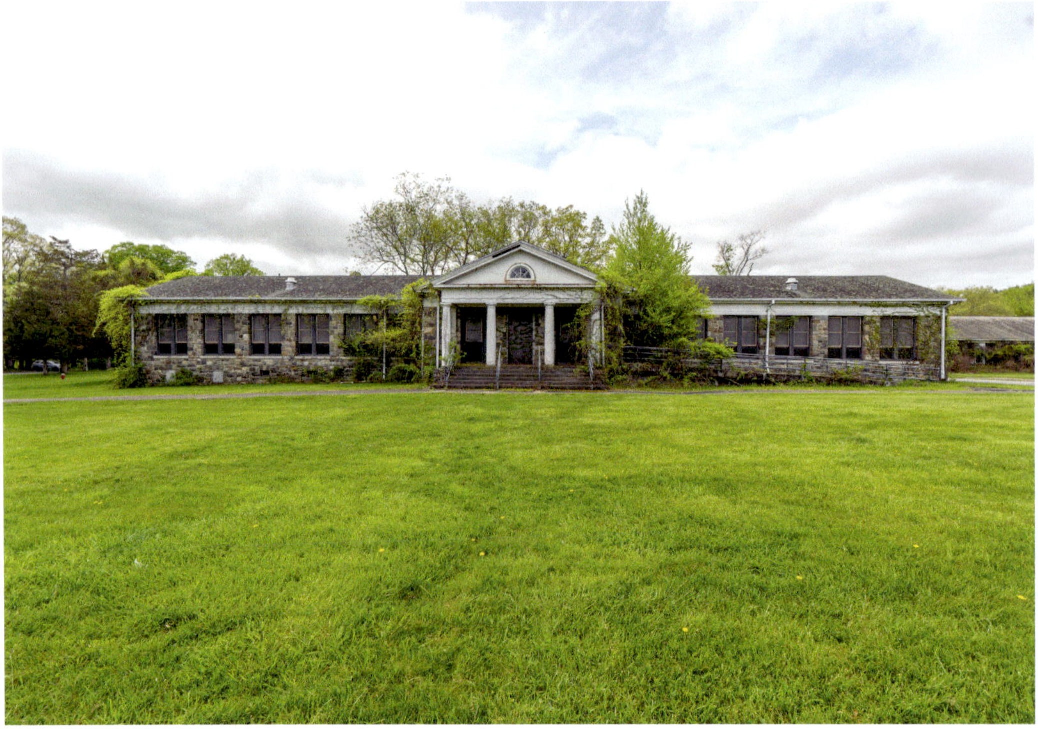

The buildings were constructed in a neoclassical style, designed to resemble Monticello.

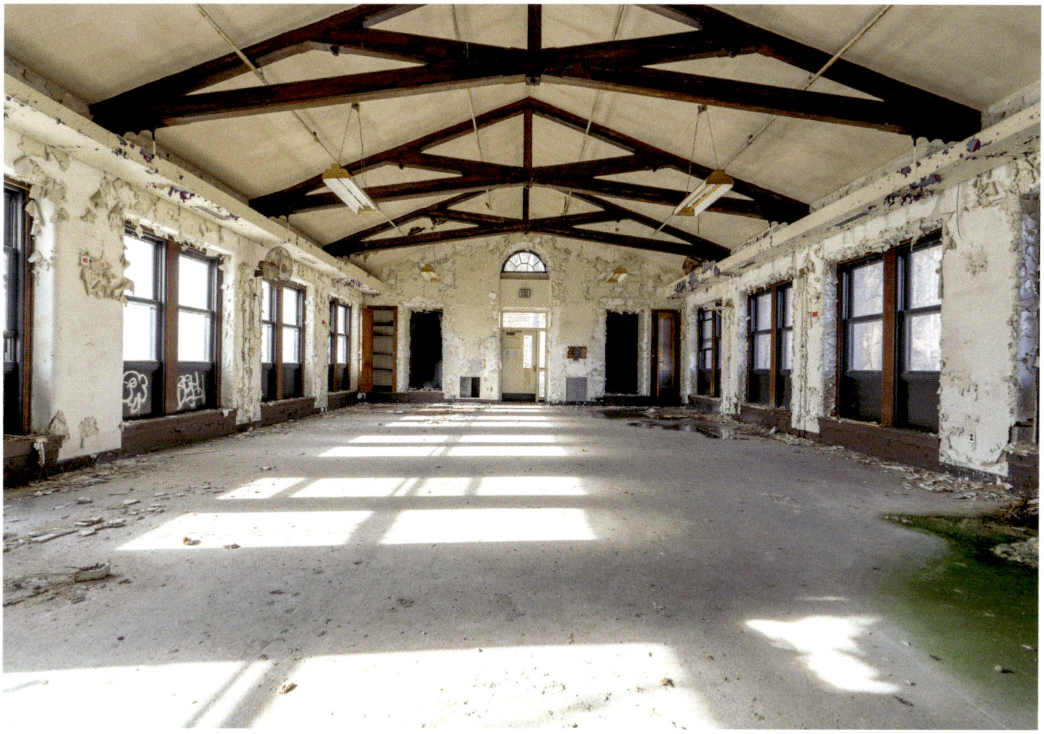

Each cottage could house up to 70 patients with three open-air wards.

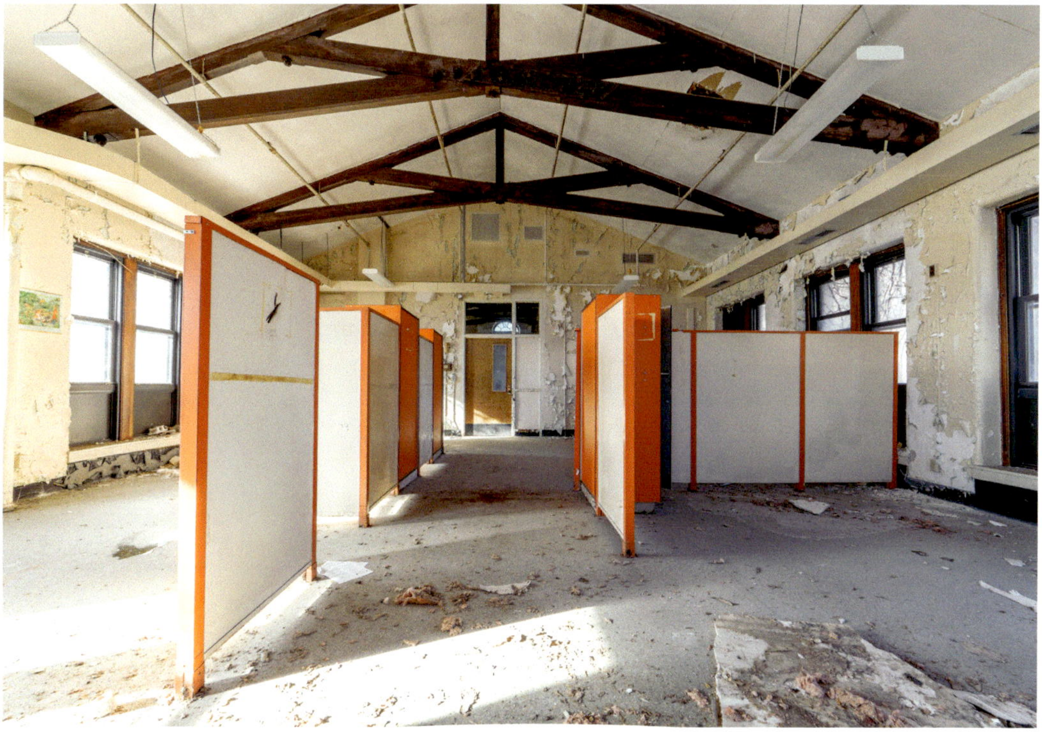

In later years room dividers would help give patients more privacy.

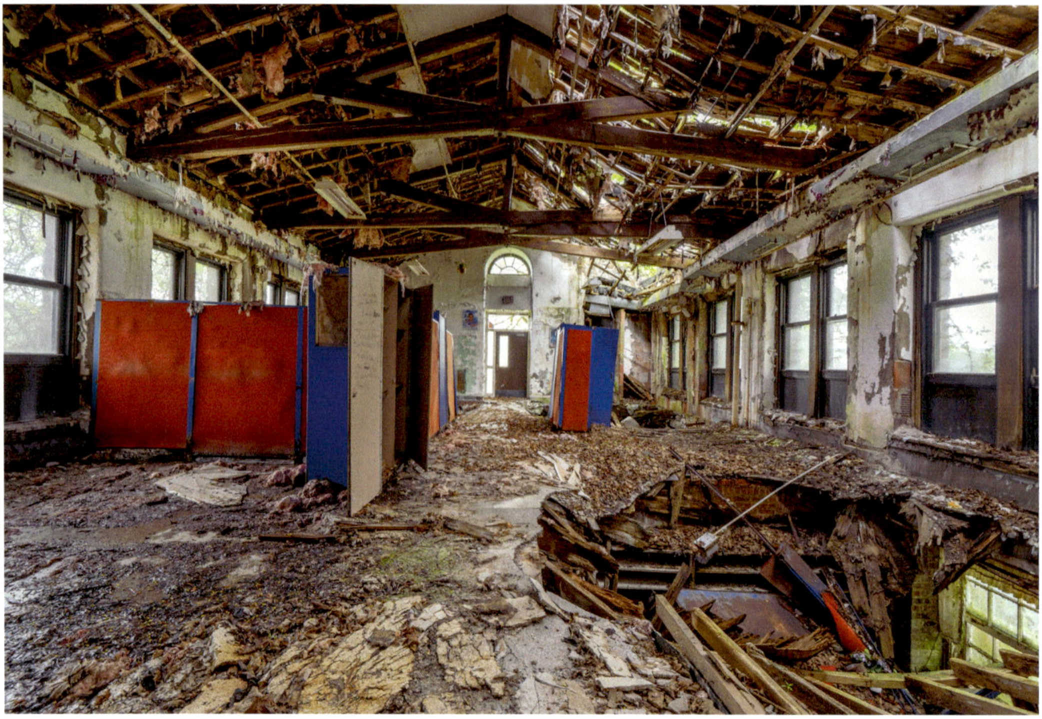

Some of the girl's group cottages were renovated into a grade school while some were left to decay.

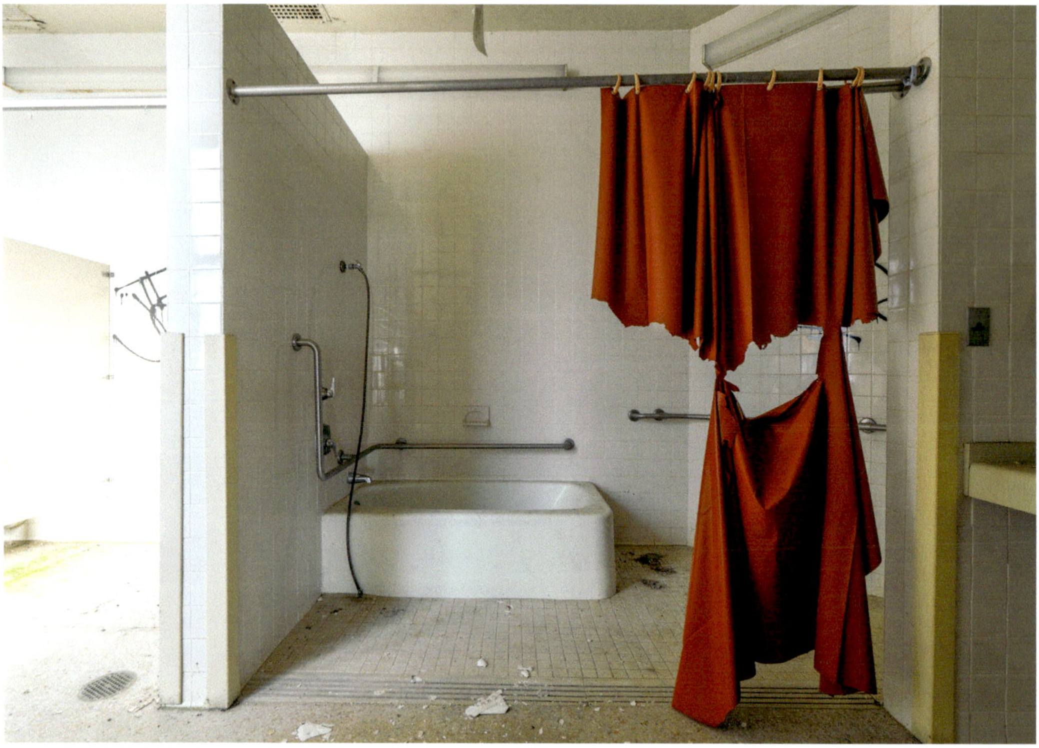

Bathrooms and showers were centralized in each cottage for functionality.

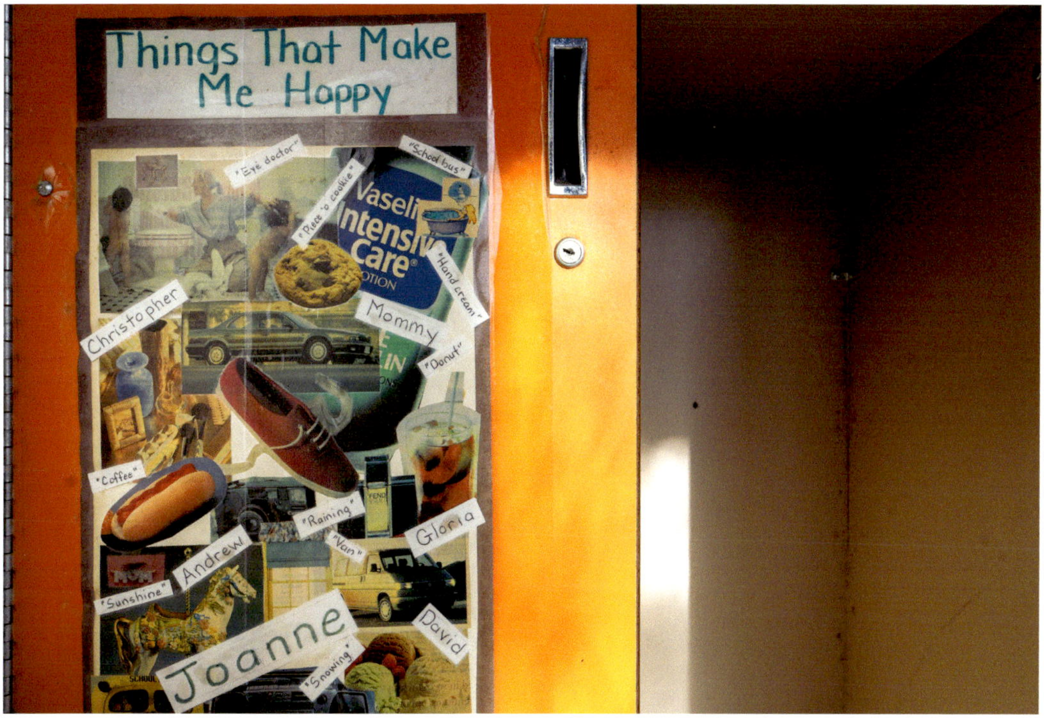

Art therapy would help patients connect or escape the world around them.

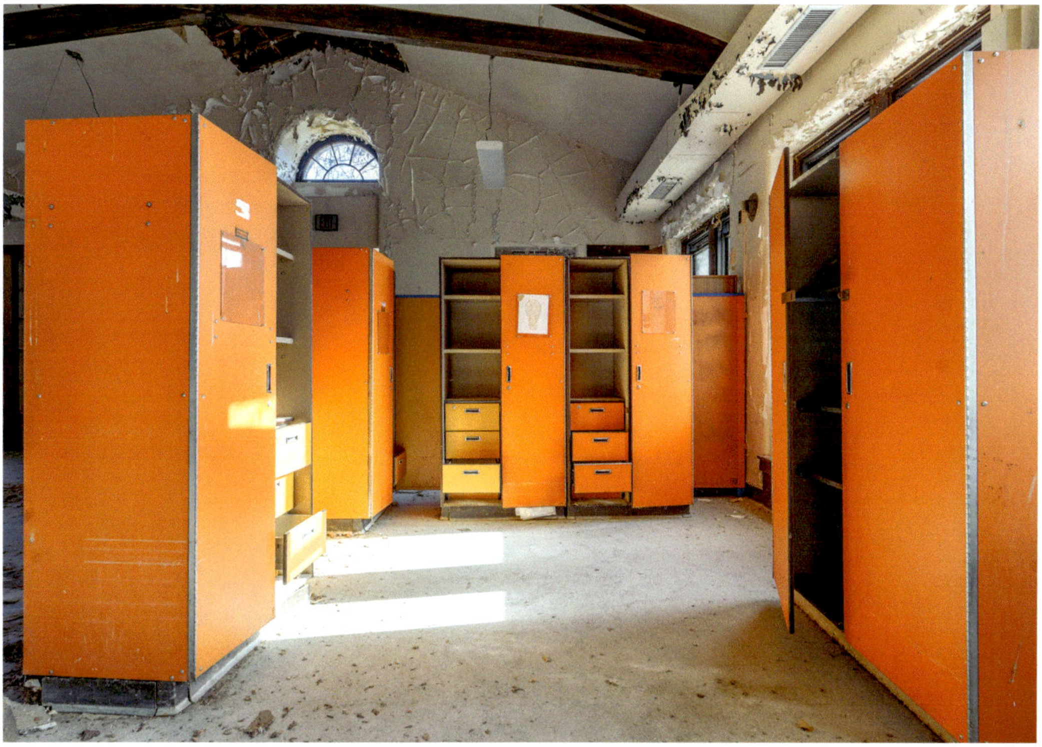

Patients would be provided with small closets in which to store their belongings.

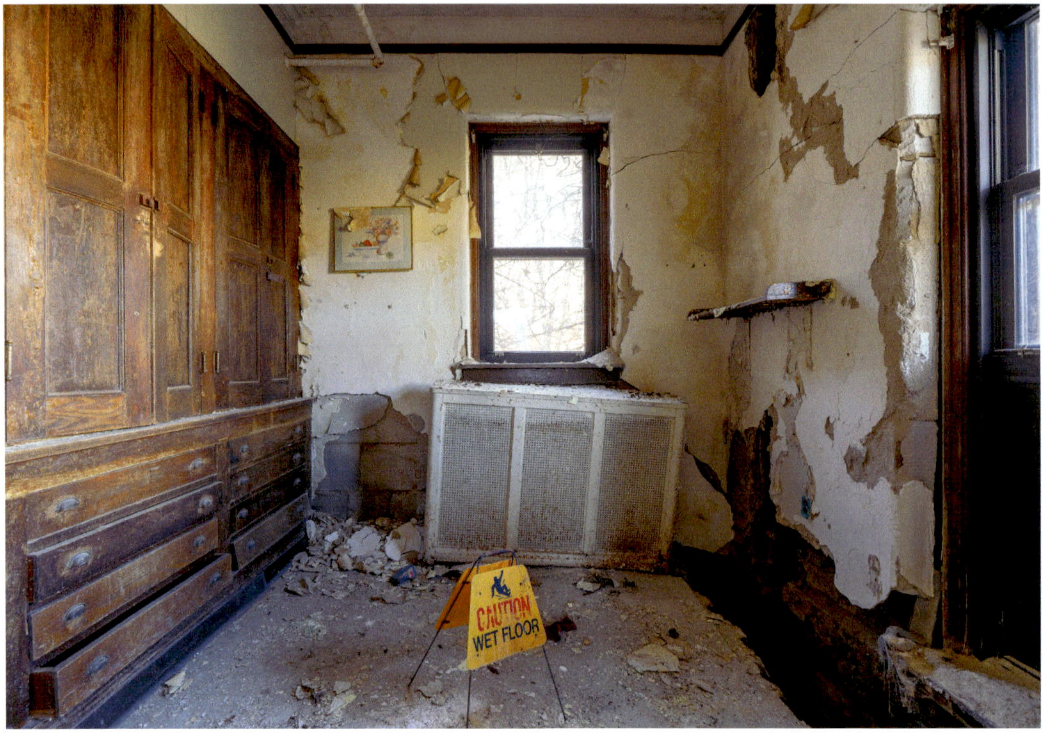

Each cottage had areas for staff to work and relax.

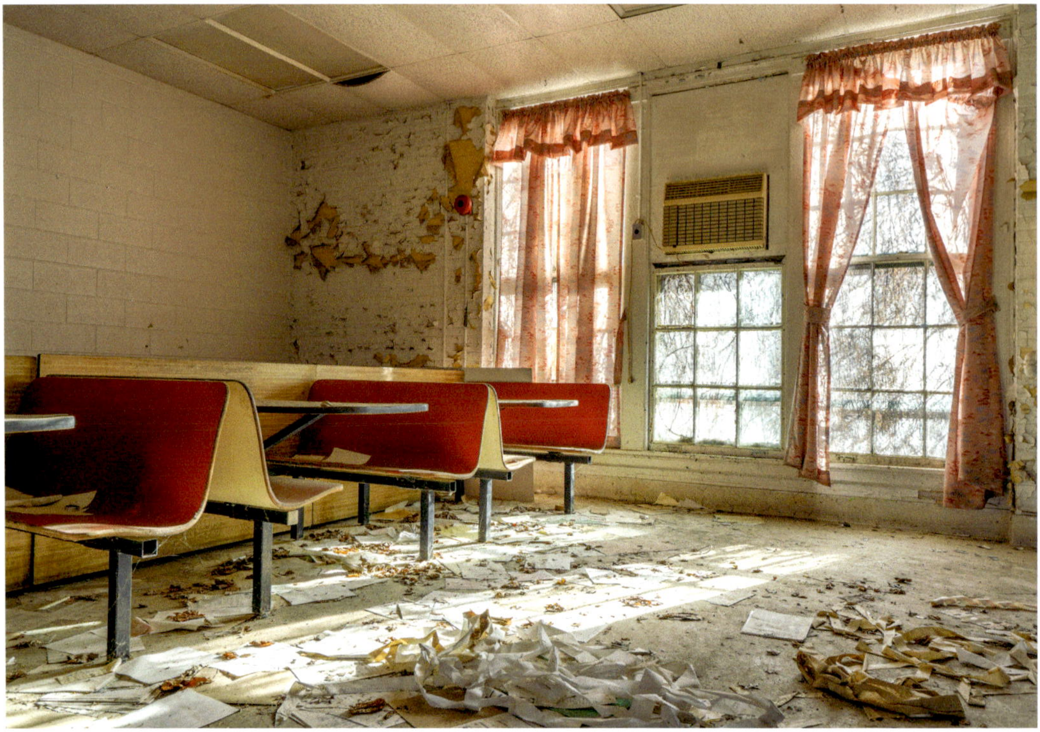

Each group of males and females had their central dining hall, which was separate from each other.

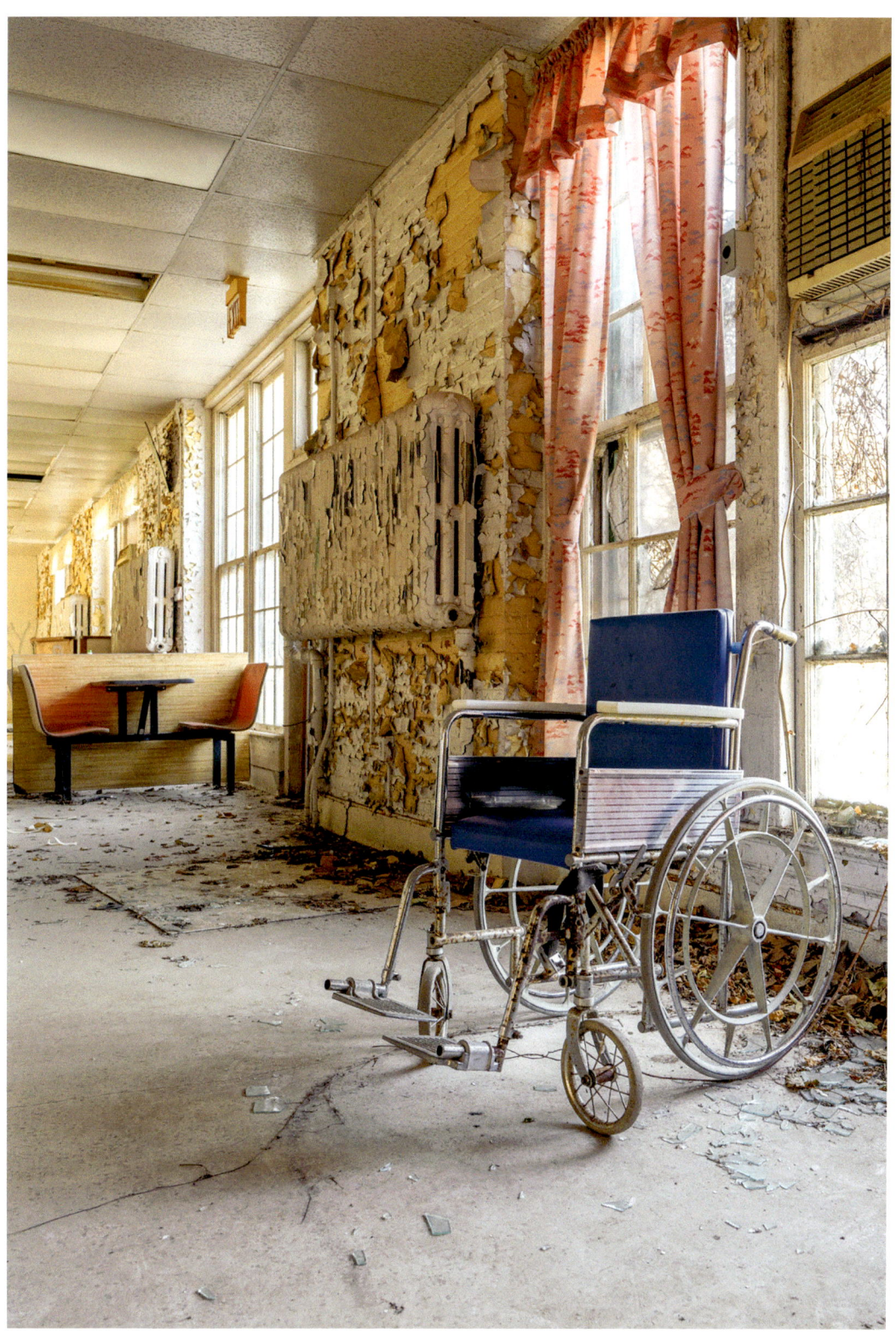
A wheelchair sits lonely in a once busy dining hall originally for female patients.

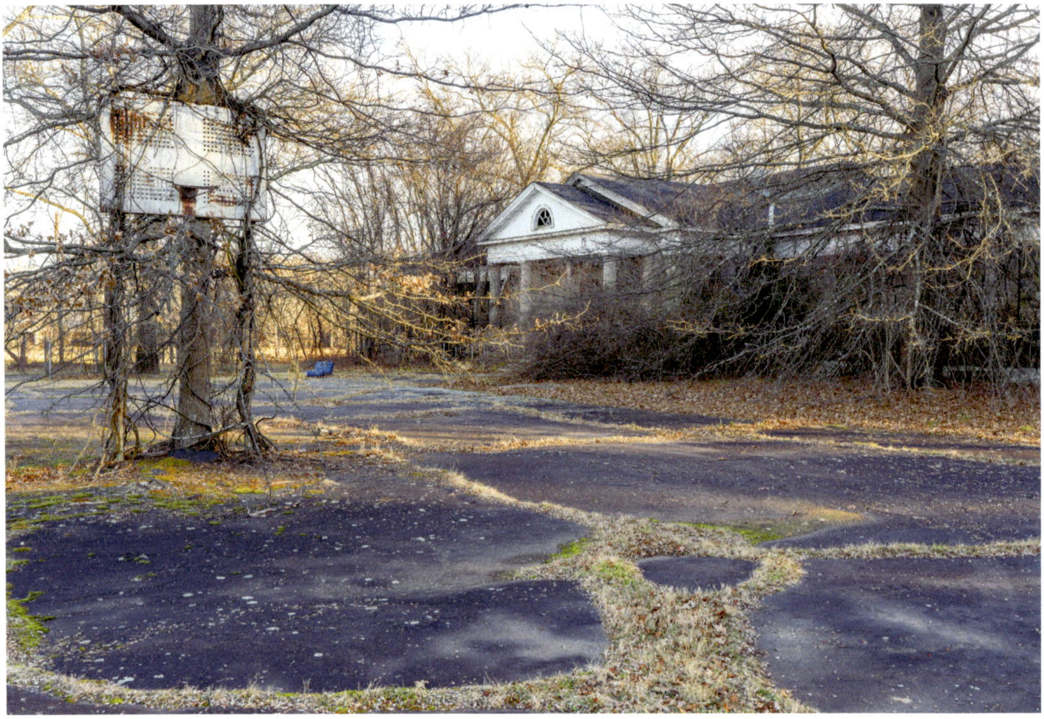

Recreational activities like basketball were common in the male group cottages.

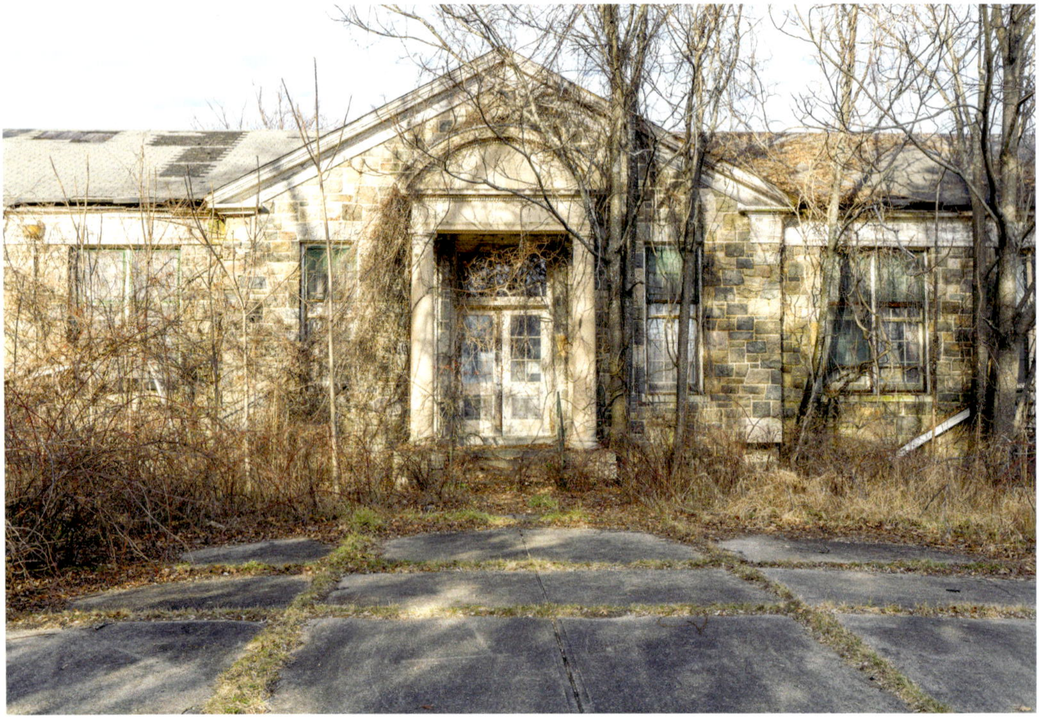

Recreation halls were helpful and essential for patients to play, read, watch TV, or socialize with one another.

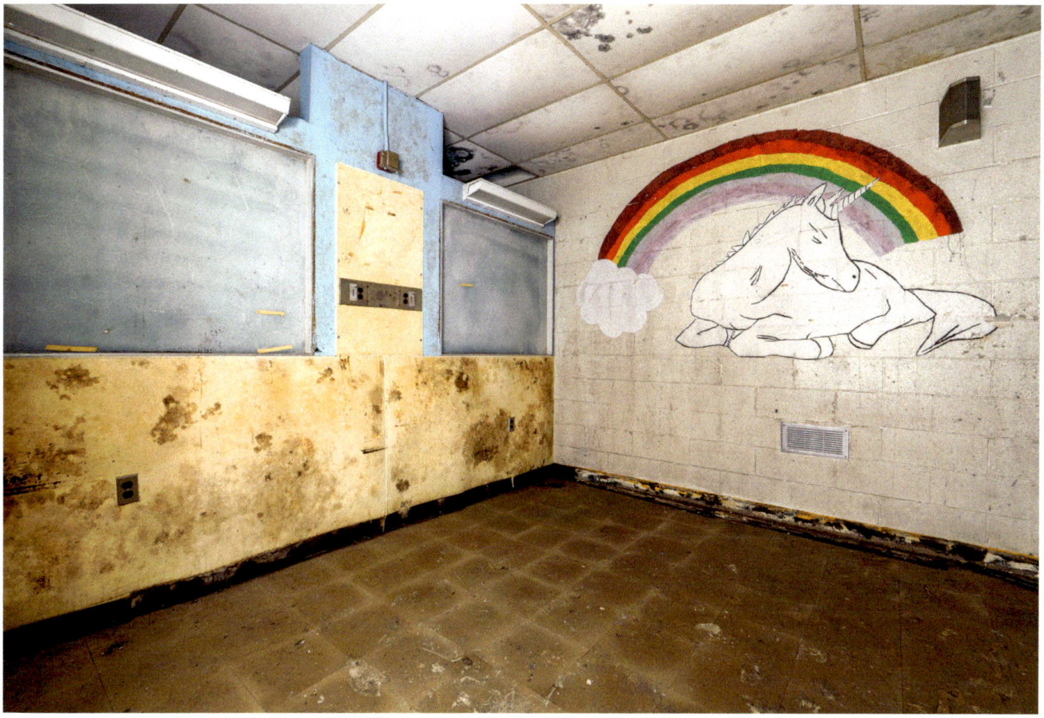

Graphic arts and murals are utilized on key surfaces in children's hospital wards to enhance well-being due to their therapeutic effects.

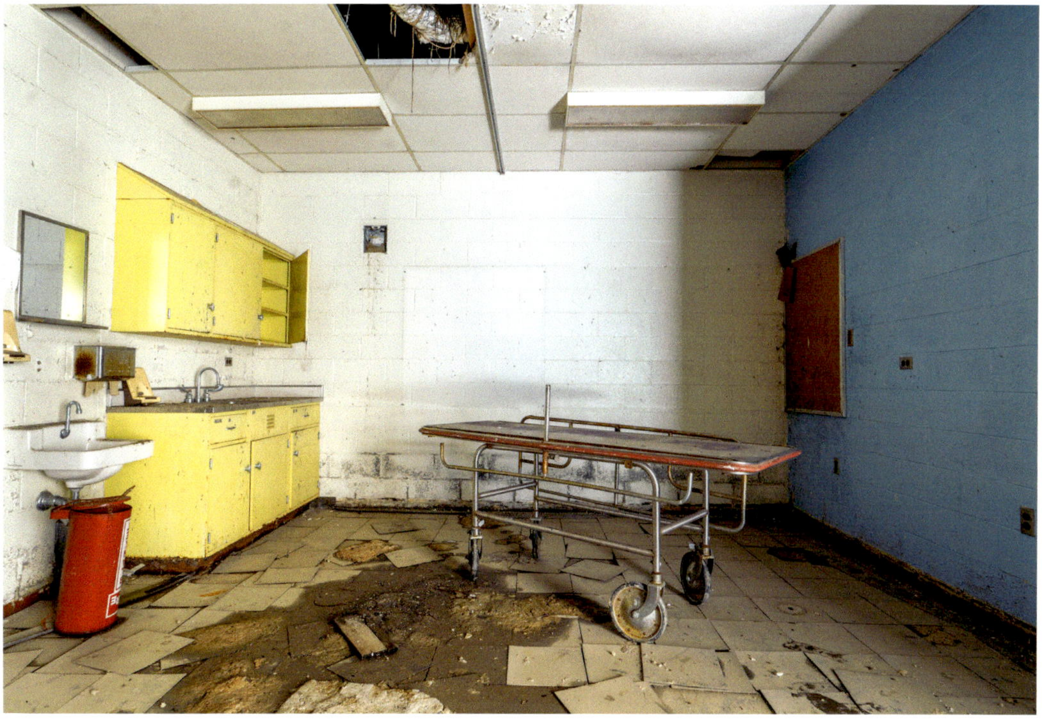

One of the earliest polio vaccines was developed and tested on children at this institution.

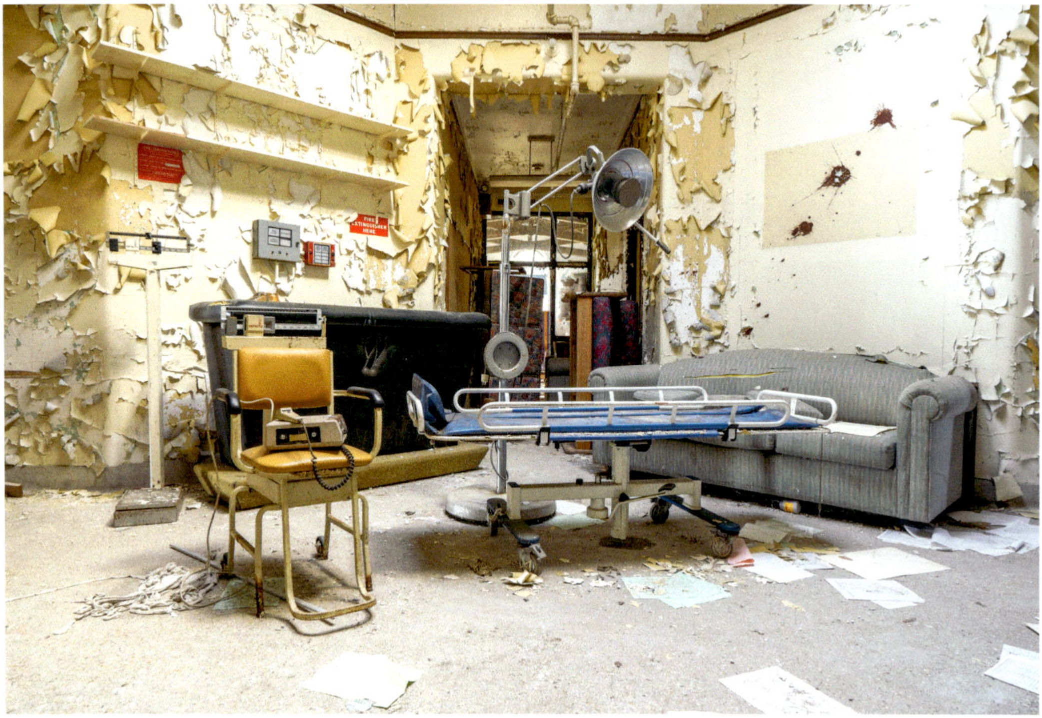

Now, unused medical equipment sits decaying in an abandoned patient cottage.

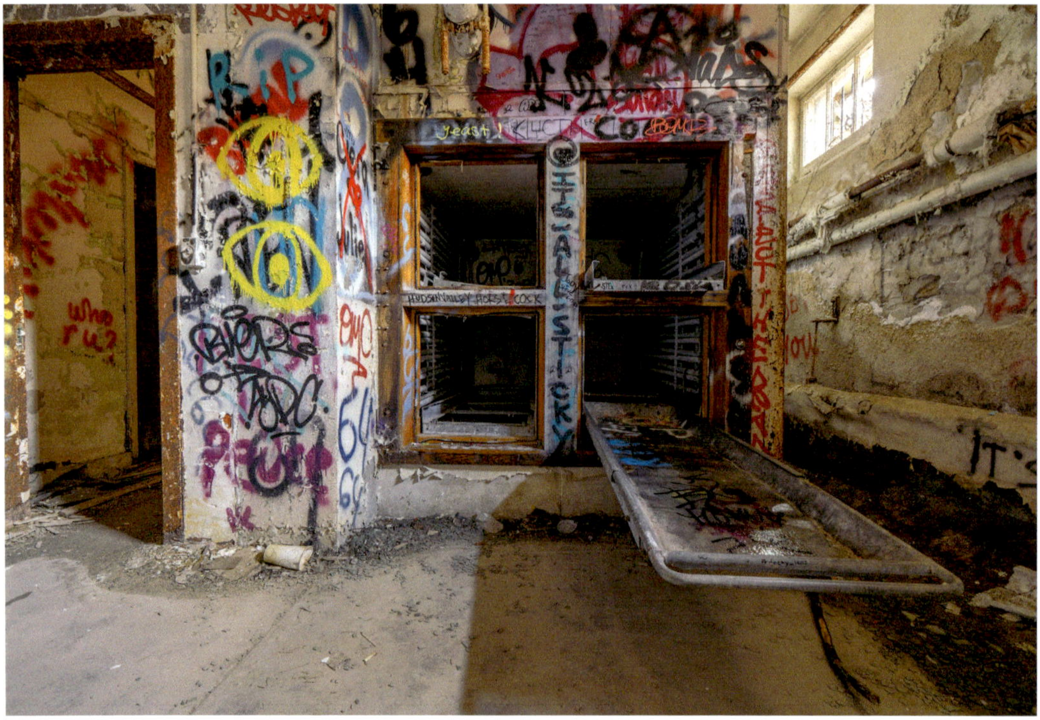

This institution comprised over 130 buildings, including staff housing, patients cottages, religious chapels, medical buildings, and two morgues.

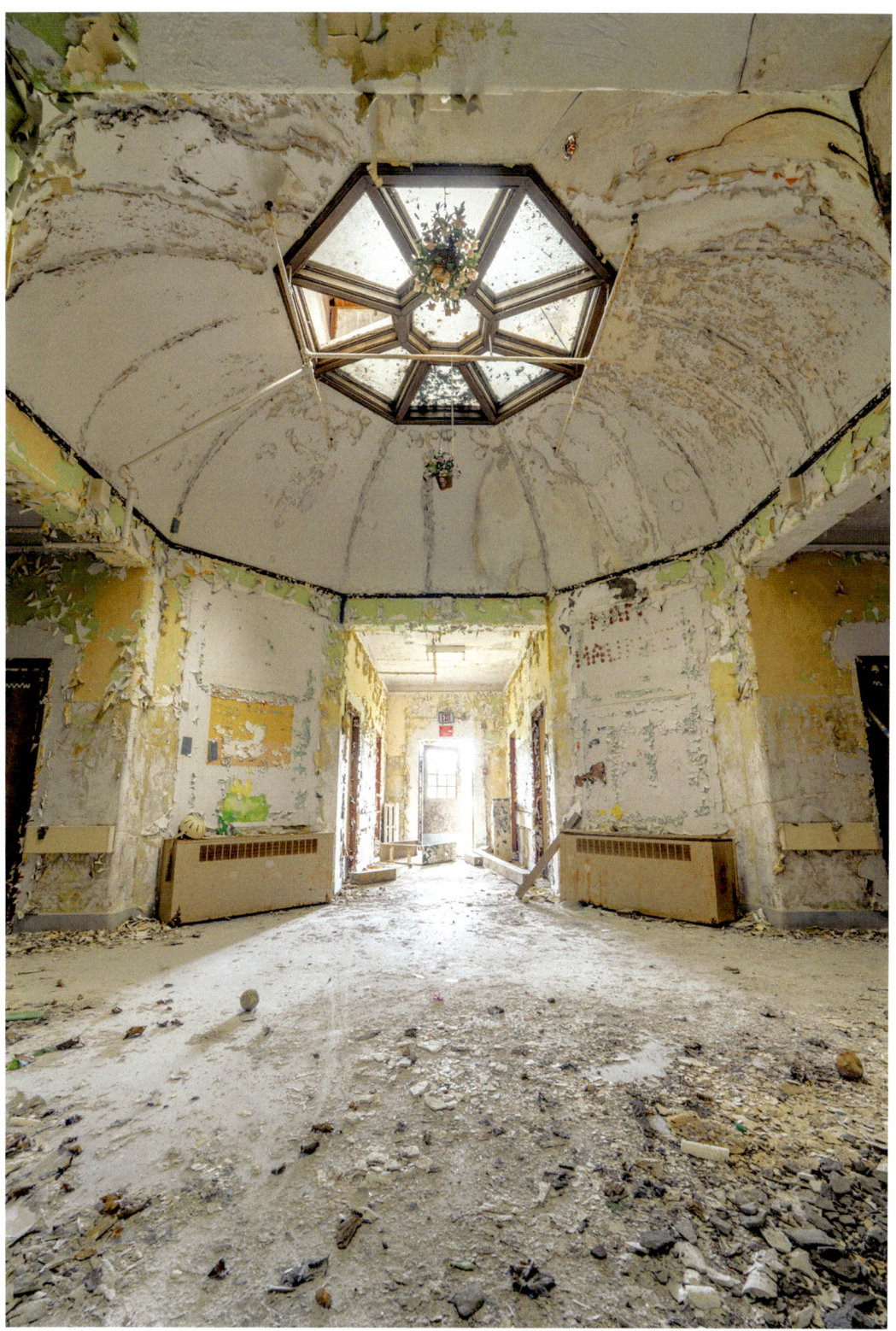

"Difficult to see; always in motion is the future."

# ABOUT THE AUTHOR

Dave Snook is a self-taught photographer from Massachusetts who has explored abandoned asylums since the early 2000s. His passion for mental health history, architecture, and preservation led him to pursue this unique underground hobby at the time. Despite the inherent dangers and uncertainties, Dave is constantly drawn back to these abandoned asylums because of the thrill and reward of capturing their beauty and documenting their rich history. To this day, he continues to document and advocate for the preservation. You can view most of his work on his website: www.Insanectuary.com.

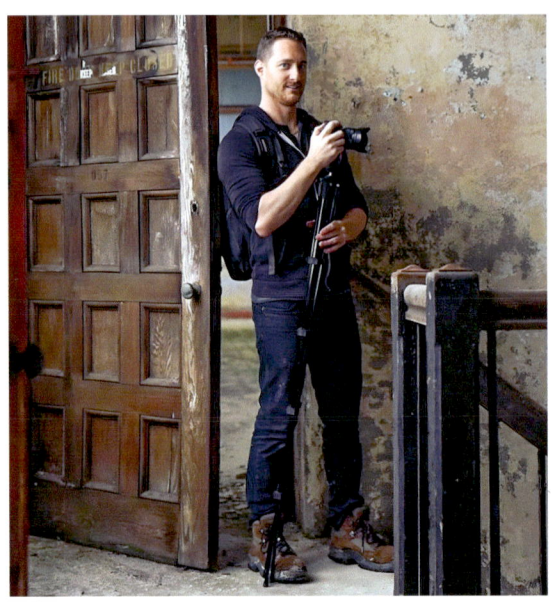